# 6th International Conference on Materials Research and Innovation 2024



IOP Conference Series: Materials Science and Engineering
Volume 1325

Bangkok, Thailand
18-19 December 2024

ISBN: 979-8-3313-1578-8
ISSN: 1757-8981

**Printed from e-media with permission by:**

Curran Associates, Inc.
57 Morehouse Lane
Red Hook, NY 12571

**Some format issues inherent in the e-media version may also appear in this print version.**



Institute of Physics
Dirac House, Temple Back
Bristol BS1 6BE UK

Phone: 44 1 17 929 7481
Fax: 44 1 17 920 0979

techtracking@iop.org

**Additional copies of this publication are available from:**

Curran Associates, Inc.
57 Morehouse Lane
Red Hook, NY 12571 USA
Phone: 845-758-0400
Fax: 845-758-2633
Email: curran@proceedings.com
Web: www.proceedings.com

# TABLE OF CONTENTS

Preface

Peer Review Statement

Energy absorption and impact protection characteristics of natural rubber foam modified with dilatant fluid ........ 1
*P Kaewpradit, C Dechwayukul*

Enhancement of thermal properties of the blends between bio-based polybenzoxazine and polydimethylsiloxane by using crosslinking agent ........ 5
*N Chaiwichian, K Hemvichian, P Lertsarawut, R Rimdusit, S Tiptipakorn*

Effects of molecular weight of polycaprolactone on the thermal properties of V-fa/PCL blends added with cross-linking agent ........ 9
*N Chaipakdee, K Hemvichian, P Lertsarawut, S Rimdusit, S Tiptipakorn*

Atomistic Simulations of 5CB Liquid Crystal on Trans- and Cis-Azo Interfaces ........ 13
*T. Fudulwatjananon, N. Nisoh, P. Nalakarn, J. Wong-Ekkabut*

Smart sensing film developed by polyvinyl alcohol/carboxymethyl cellulose doped with anthocyanin extract as a natural pH indicator ........ 17
*P Rieagnam, P Jaisabai, K Wongkhan, R Jitchati*

Development of polylactic acid composite enhance with cassava pulp and cassava strach ........ 25
*B Ketnapha, F Saranyakan, W Kittiya, J Rukkiat*

Development of the thermoplastic vulcanizate material between bromobutyl rubber and polypropylene: compatibility study ........ 32
*Siriwan Klinjan, Supansa Khunkayan, Peerut Suwanpimolkul, Supitta Suethao, Wirasak Smitthipong*

QM/MM/3D-RISM study of an electronic structure change of benzimidazole derivative induced by molecular recognition of cucurbit[7]uril ........ 37
*Kodai Kanemaru, Keisuke Shiroshita, Haruyuki Nakano, Norio Yoshida*

Enhance fire behavior of rigid polyurethane foam by intumescent flame retardants ........ 41
*V Peasuwan, W Keawwattana, H Paeprapa, M Raoarun, K Hancharoen*

Findings the proper high styrene resin/accelerator/antioxidant ratios to natural rubber/reclaimed rubber composites: Cure characteristics, mechanical properties and weathering resistance ........ 46
*P Boochatum, W Mingvanish, Y Nakaramontri*

Effect of vegetable oil on the properties of rubber compound ........ 54
*S Suethao, S Phongphanphanee, J Wong-ekkabut, H Yangthong, W Smitthipong*

**Author Index**

6TH INTERNATIONAL
CONFERENCE ON MATERIALS
RESEARCH AND INNOVATION
(6TH ICMARI 2024)
18TH - 19TH DECEMBER 2024
AT CENTARA GRAND AT CENTRAL PLAZA LADPRAO, BANGKOK, THAILAND
IOP PROCEEDING
Register
CONTACT INFORMATION:
Phone: +66623979554
www.icmari.sci.ku.ac.th
Email: icmari.secretary@gmail.com
Poster/Oral presentation
Online/ Onsite
Wear mask
Use soap & antibacteria gel
Keep your distance
HORIBA Scientific
CERNTEK
PMUC
LMS
CG SCIENTIFIC
CT
CG ENGINEERING CO.,LTD.
Publish in English
TRT
SK POLYMER
METTLER TOLEDO
Sci Spec
TREASURE PRODUCTS
SIAM UNITED RUBBER

IOP Conf. Series: Materials Science and Engineering **1325** (2025) 011001 doi:10.1088/1757-899X/1325/1/011001

# Preface

The $6^{th}$ International Conference on Materials Research and Innovation ($6^{th}$ ICMARI) was held on $18^{th}$ – $19^{th}$ December 2024 in Bangkok, Thailand. The conference was conducted in frontier research on materials research and Innovation including Natural Rubber Science and Engineering, Sustainable Polymers and Biomaterials, and Special Advanced Materials. This $6^{th}$ ICMARI was announced to focused on the Bio-Circular-Green Economic Model or BCG. The BCG has been applied by the research community and promoted by the Thai government as a new economic model for inclusive and sustainable growth, this BCG model also conforms with the UN Sustainable Development Goals (SDGs). The $6^{th}$ ICMARI was a great opportunity for all participants to exchange knowledge and strengthen the research collaboration not only in ASEAN but also around the world. We hoped that the participants are fruitful by plenary and invited speakers from novel knowledge to new applications in the field of materials research and innovation.

Thank you all for participating,

ICMARI 2024 Organizing Committee

IOP Conf. Series: Materials Science and Engineering **1325** (2025) 011001 doi:10.1088/1757-899X/1325/1/011001

## Organizing Committee

| | | |
|---|---|---|
| Assoc. Prof. Wirasak | Smitthipong | (Chairman) |

## Committee (Inside KU)

| | |
|---|---|
| Asst. Prof. Saree | Phongphanphanee |
| Asst. Prof. Nattasamon | Petchsang |
| Assoc. Prof. Decha | Dechtrirat |
| Assoc. Prof. Pongthep | Prajongtat |
| Assoc. Prof. Sutee | Boonchui |
| Assoc. Prof. Jirasak | Wong-ekkabut |
| Asst. Prof. Chomdao | Sinthuvanich |
| Asst. Prof. Pichamon | Kiatwuthinon |
| Asst. Prof. Wirunya | Keawwattana |
| Assoc. Prof. Kiadtisak | Saenboonruang |
| Assoc. Prof. Nanthiya | Hansupalak |
| Assoc. Prof. Prakit | Sukyai |
| Asst. Prof. Parames | Kamhangritirong |
| Assoc. Prof. Chanapa | Kongmark |
| Assoc. Prof. Thidarat | Supasai |
| Assoc. Prof. Chantiga | Choochottiros |
| Asst. Prof. Pimsiree | Suwanna |
| Dr. Supitta | Suethao |
| Dr. Nattawut | Yuntawattana |
| Dr. Hassarutai | Yangthong |
| Ms. Siriwan | Nerdnoi |

## Committee (outside KU)

| | |
|---|---|
| Mr. Payab | Namprasert |
| Mrs. Pannee | Jirabunpaisit |
| Assoc. Prof. Nitinart | Saetung |
| Assoc. Prof. Netnapid | Tantemsapya |
| Asst. Prof. Wiboon | Lertwimolnun |

## Secretary team

| | |
|---|---|
| Dr. Hassarutai | Yangthong |
| Mr. Kwanchai | Buaksuntear |
| Ms. Phakamat | Lim-arun |
| Ms. Haruthai | Klommueang |
| Ms. Siriwan | Klinjan |

## Editors of Proceeding

| | | |
|---|---|---|
| Assoc. Prof. Wirasak | Smitthipong | Chair Editor |
| Asst. Prof. Saree | Phongphanphanee | Editor |
| Assoc. Prof. Jirasak | Wong-ekkabut | Editor |

IOP Conf. Series: Materials Science and Engineering **1325** (2025) 011001 doi:10.1088/1757-899X/1325/1/011001

## International Technical Committee

| | | | |
|---|---|---|---|
| Prof. Sadhan C. | Jana | University of Akron | USA |
| Prof. Arkadiusz | Chworos | Polish Academy of Sciences | Poland |
| Assoc. Prof. Darshill U | Shah | University of Cambridge | UK |
| Prof. Hongbo | Zeng | University of Alberta | Canada |
| Assoc. Prof. Karine | Mougin | Institut de Science des Materiaux de Mulhouse | France |
| Prof. Chen | Soo Kien | Universiti Putra Malaysia | Malaysia |
| Prof. Katsura | Nishiyama | Meijo University | Japan |
| Prof. Norio | Yoshida | Nagoya University | Japan |
| Asst. Prof. Kentaro | Matsumoto | Nagoya University | Japan |
| Prof. Hisashi | Okumura | Institute fof Molecular Science | Japan |
| Assoc. Prof. Wirasak | Smitthipong | Kasetsart University | Thailand |
| Asst. Prof. Saree | Phongphanphanee | Kasetsart University | Thailand |
| Dr. Supitta | Suethao | Kasetsart University | Thailand |
| Dr. Nattawut | Yuntawattana | Kasetsart University | Thailand |
| Dr. Hassarutai | Yangthong | Kasetsart University | Thailand |
| Prof. Jitladda | Sakdapipanich | Mahidol University | Thailand |
| Assoc. Prof. Nitinart | Saetung | Prince of Songkla University | Thailand |
| Assoc. Prof. Netnapid | Tantemsapya | Suranaree University of Technology | Thailand |

IOP Conf. Series: Materials Science and Engineering **1325** (2025) 011001 doi:10.1088/1757-899X/1325/1/011001

# Peer review statement

All papers published in this volume of *IOP Conference Series: Materials Science and Engineering* have been peer reviewed through processes administered by the proceedings Editors. Reviews were conducted by expert referees to the professional and scientific standards expected of a proceedings journal published by IOP Publishing.

IOP Conf. Series: Materials Science and Engineering **1325** (2025) 011002 doi:10.1088/1757-899X/1325/1/011002

# Peer Review Statement

All papers published in this volume have been reviewed through processes administered by the Editors. Reviews were conducted by expert referees to the professional and scientific standards expected of a proceedings journal published by IOP Publishing.

- **Type of peer review:** Single Anonymous
- **Conference submission management system:** Morressier
- **Number of submissions received:** 17
- **Number of submissions sent for review:** 17
- **Number of submissions accepted:** 17
- **Acceptance Rate (Submissions Accepted / Submissions Received × 100):** 100
- **Average number of reviews per paper:** 2
- **Total number of reviewers involved:** 20
- **Contact person for queries:**
  **Name:** Hassarutai Yangthong
  **Email:** fscihty@ku.ac.th
  **Affiliation:** Kasetsart University - Kasetsart University Faculty of Social Sciences

IOP Conf. Series: Materials Science and Engineering **1325** (2025) 012001 doi:10.1088/1757-899X/1325/1/012001

# Energy absorption and impact protection characteristics of natural rubber foam modified with dilatant fluid

**P Kaewpradit[1]*, and C Dechwayukul[2]**

[1] Department of Chemical Engineering, Faculty of Engineering, Prince of Songkla University, Bangkok 90110, Thailand
[2] Department of Mechanical Engineering, Faculty of Engineering, Prince of Songkla University, Bangkok 90110, Thailand

*Corresponding author: pornsiri.k@psu.ac.th

**Abstract.** This study investigates the dynamic mechanical behavior and impact protection performance of natural rubber foam modified with varying concentrations of a developed dilatant fluid (0, 5, 10, and 15 phr), which exhibits shear thickening behavior. The energy absorption capability was evaluated through hysteresis loop analysis, while impact protection performance was assessed using drop tests. The hysteresis results demonstrated that the incorporation of 5 phr dilatant fluid significantly enhanced the energy absorption capacity of the foam, exhibiting the largest hysteresis loop area compared to other concentrations. Samples containing 0, 10, and 15 phr dilatant fluid showed insignificant differences in energy absorption behavior. Notably, the drop test analysis revealed that increasing dilatant fluid content led to a reduction in peak force, indicating improved impact protection properties. These findings suggest that while 5 phr dilatant fluid provides optimal energy absorption under cyclic loading, higher concentrations of dilatant fluid contribute to better impact force attenuation. This study provides valuable insights for developing enhanced impact protection materials for various applications requiring both energy absorption and force transmission reduction capabilities.

## 1. Introduction

Impact protection materials play a crucial role in various applications, from personal protective equipment to automotive safety systems. Among these materials, natural rubber (NR) foam has gained significant attention due to its excellent elasticity, durability, and cost-effectiveness [1]. However, the growing demand for enhanced impact protection has led researchers to explore various modifications to improve its energy absorption and impact attenuation capabilities [2-3]. One promising approach to enhance the performance of NR foam is the incorporation of dilatant fluid, also known as shear thickening fluids (STFs). These STFs exhibit unique rheological properties, where their viscosity increases with applied shear rate, making them particularly effective in impact protection applications. The combination of NR foam's inherent elasticity with the strain-rate dependent behaviour of the fluids presents an opportunity to develop materials with superior impact protection characteristics compared to neat NR foam [4-6].

The concentration of dilatant fluid plays a critical role in determining both the energy absorption capacity under cyclic loading and the force attenuation during impact events. This study then focuses on investigating the effects of varying dilatant fluid concentrations on the dynamic mechanical behaviour and impact protection performance of natural rubber foam.

## 2. Experimental

The experimental investigation was conducted using natural rubber (NR) as the base material. The rubber compounds were prepared at different amount of dilatant fluid, which exhibits shear thickening behavior (Table 1) using a Banbury internal mixer (Thermo Scientific HAAKE), then vulcanized using a thermal compression moulding machine at 150°C.

Compression set was assessed utilizing a compression set tester (Gotech brand, model GT-7049) at 50%strain, 70°C for 22 hr in accordance with the ASTM D1056-07. Hardness measurements were performed using a Durometer (Bareiss brand, model BS61-II), following the ASTM D2240-5. The hysteresis loop testing was conducted at a compression speed of 500 mm/min, applying a maximum strain of 65%, with three specimens tested (three cycles each). For the drop test, the initial impact energy was set at 5 Joules (an impact mass of 8.5 kg), following the ASTM F 1614-99.

**Table 1.** Formulas of NR foam (parts per hundred of rubber).

| Formula | STR20 | Silica | Clay | BW | Dilatant |
|---|---|---|---|---|---|
| 1 | 100 | 20 | 10 | 3 | 0 |
| 2 | 100 | 20 | 10 | 3 | 5 |
| 3 | 100 | 20 | 10 | 3 | 10 |
| 4 | 100 | 20 | 10 | 3 | 15 |

## 3. Results and discussion

The results (Figure 1) demonstrated that the incorporation of dilatant fluid significantly influences the material's energy absorption capacity. The NR foam (size of 15 cm × 15 cm × thickness 1 cm) containing 5 phr dilatant fluid exhibited superior energy absorption properties at low impact speed, since of the largest hysteresis loop area (175±4 kPa) and highest median $F_{max}$ (1,758±52 N). This performance suggests a critical concentration threshold where the dilatant fluid effectively enhances the rubber network's ability to store and dissipate energy. In contrast, the control sample (0 phr) showed moderate energy absorption with a loop area of 110±1 kPa, while higher concentrations demonstrated reduced energy absorption capabilities with loop areas of 94±10 and 69±4 kPa, respectively.

**Table 2.** Effect of dilatant fluid on properties of NR foam.

| Dilatant (phr) | ML(1+4) @100°C | Density (g/cm3) | Hardness (Shore OO) | Compression set (%) | Hysteresis | | Peak force (N) |
|---|---|---|---|---|---|---|---|
| | | | | | Median $F_{max}$ (N) | Loop (kPa) | |
| 0 | 30.5±0.0 | 0.722±0.00 | 77.3±1.5 | 57.8±1.5 | 1,430±155 | 110±1 | 1,876±10 |
| 5 | 26.5±0.1 | 0.705±0.00 | 82.4±0.9 | 52.1±2.6 | 1,758±52 | 175±4 | 1,900±44 |
| 10 | 25.6±1.2 | 0.710±0.01 | 81.9±1.5 | 74.3±5.6 | 1,171±31 | 94±10 | 1,740±16 |
| 15 | 25.3±1.7 | 0.673±0.01 | 83.2±1.6 | 68.8±5.0 | 896 ± 69 | 69±4 | 1,650±16 |

The decreasing density trend (0.722 to 0.673 g/cm³) and Mooney viscosity reduction (30.5 to 25.3 ML(1+4)) with increasing dilatant fluid content suggest changes in the material's molecular mobility and network structure. At 5 phr concentration, an optimal balance is achieved between rubber elasticity and fluid mobility at low compressive speed, enabling efficient energy dissipation through molecular chain alignment during deformation while maintaining structural integrity. However, beyond this concentration, the excess fluid appears to interfere with the rubber network's energy storage and dissipation mechanisms [7].

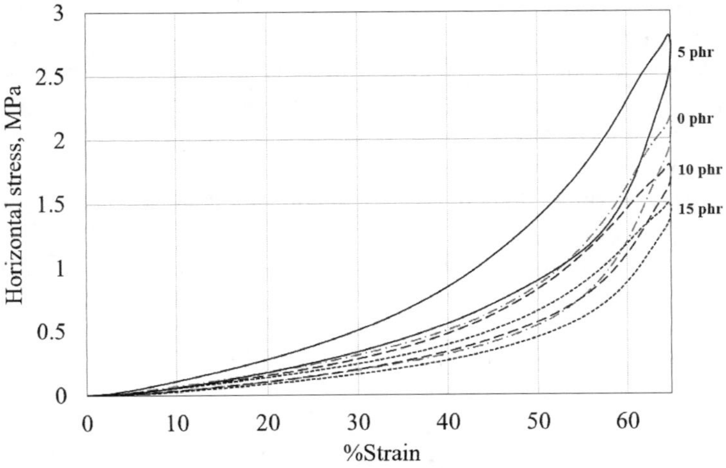


**Figure 1**. Hysteresis curves of the NR foam under cyclic loading (3th loop)

The impact protection performance of dilatant fluid-modified NR foam was systematically evaluated through drop tests at 5J impact energy, revealing a clear correlation between the fluid concentration and force attenuation capabilities. The results demonstrated a consistent trend of decreasing peak force with increasing the fluid content, from 1,876±10 N to 1,650±16 N, representing approximately 12% reduction in peak force. Polymer foam can absorb an amount of energy in the plateau region through cell wall bending and buckling. The decreasing density trend with increasing dilatant fluid content suggests that the fluid creates a more open cellular structure in the foam [8-10]. The increase in hardness despite lower density indicates that the fluid contributes to the material's resistance to surface deformation. The reduction in Mooney viscosity implies that the fluid acts as a processing aid, potentially improving molecular chain mobility. The fluid may absorb more energy than neat foam in low-velocity impact tests, with results similar to those reported in [11].

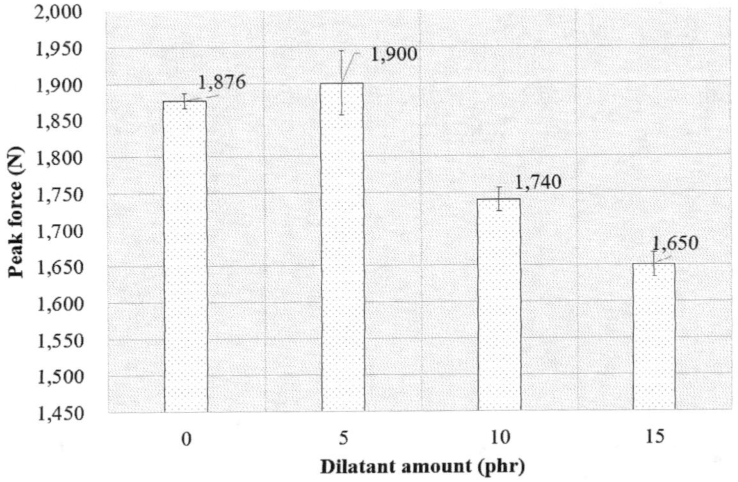


Fig. 2. Peak force at different amounts of dilatant fluid (5J impact energy).

## 4. Conclusions

This study demonstrates the relationship between dilatant fluid concentration and the performance characteristics of natural rubber (NR) foam. The results reveal distinct optimal concentrations for different performance requirements: 5 phr dilatant fluid exhibits superior energy absorption capacity under cyclic loading (low speed, ≅ 0.008 m/s) with the highest hysteresis loop area (175±4 kPa), while higher concentrations (15 phr) provide better impact protection with the lowest peak force (1,650±16 N) under 5J impact energy (impact speed ≅ 1.08 m/s). The addition of dilatant fluid systematically reduces material density and Mooney viscosity, while increasing hardness and compression set. This trade-off between energy absorption and impact protection capabilities can be attributed to the competing mechanisms of molecular chain mobility and fluid-induced stress distribution. These findings provide valuable insights for tailoring material properties to specific applications. The foam with 5 phr dilatant fluid (rate dependent behaviour) is recommended for applications requiring cyclic loading resistance, while the foam with 15 phr fluid is more suitable for higher impact protection scenarios. This understanding enables the strategic development of enhanced protective materials with optimized performance characteristics for diverse applications.

## Acknowledgement

This work was funded by National Research Council of Thailand. The authors would like to thank Faculty of Engineering, and International College, Prince of Songkla University for all their support.

## References

[1] Rostami-Tapeh-Esmaeil E, Vahidifar A, Esmizadeh E and Rodrigue D 2021 *Polym.* **13**(10) 1565.
[2] Silva R.M., Rodrigues J.L., Pinto V.V., Ferreira M.J., Russo R., Pereira C.M. 2009 *Polym. Test.* **28**(6) 642.
[3] Tomin M and Kmetty A 2022 *J Appl Polym Sci.* **139** e51714.
[4] Kaewpradit P, Kongchoo A, Chonlathan P, Lehman N and Kalkornsurapranee E 2020 *SPE Polym.* **2** 217.
[5] Kaewpradit P, Uthaipan N and Dechwayukul C 2023 *IOP Conf. Ser.: Mater. Sci. Eng.* **1280** 012023.
[6] Wang S, Zhang W, Gao J, Gao D and He C 2024 *Int J Adv Manuf Technol* **133** 1973.
[7] Fu K, Wang H, Zhang Y.X., Ye L, Escobedo J P., Hazell P J., Friedrich K and Dai 2020 S *Int. J. Impact Eng.* **139** 103525.
[8] Williams M.K., Weiser E.S., Fesmire J.E., Grimsley B.W., Smith T.M., Brenner J.R. and Nelson G.L 2005 *Polym. Adv. Technol.* **16** 167.
[9] Tomin M, Torok D, Paszthy T and Kmetty A 2023 *Polym. Test.* **122** 108014.
[10] Liu H, Fu K, Cui X, Zhu H and Yang B 2023 *Polym.* **15** 2238.
[11] Fu K, Wang H, Chang L, Foley M., Friedrich K and Ye L 2018 *Compos Sci & Technol.* **165** 74.

IOP Conf. Series: Materials Science and Engineering **1325** (2025) 012002 doi:10.1088/1757-899X/1325/1/012002

# Enhancement of thermal properties of the blends between bio-based polybenzoxazine and polydimethylsiloxane by using crosslinking agent

**N Chaiwichian[1], K Hemvichian[1], P Lertsarawut[2], R Rimdusit[3], and S Tiptipakorn[1,*]**

[1] Department of Physical and Material Sciences, Faculty of Liberal Arts and Science, Kasetsart University, Nakhon Pathom, 73140, Thailand.
[2] Thailand Institute of Nuclear Technology (Public Organization), Ongkarak District, Nakornnayok 26120, Thailand.
[3] Center of Excellence in polymer Materials for Medical Practice Devices, Department of Chemical Engineering, Faculty of Engineering, Chulalongkorn University, Bangkok, 10330, Thailand.

* Corresponding author: faassntk@ku.ac.th

**Abstract.** In this research, the polymer blends between bio-based polybenzoxazine (V-fa) and polydimethylsiloxane (PDMS) were synthesized at V-fa:PDMS mass ratios of 1:1, 2:3, 1:2, and 1:4. γ-Aminopropyltriethoxysilane (KH550) was used as crosslinking agent. The chemical structure was characterized using Fourier Transform Infrared Spectroscopy (FT-IR), which indicated the formation of specific bonds between poly(V-fa) and PDMS. The disappearance of the peak corresponding to -OH groups suggested their involvement in the coupling between two polymers. The thermal properties were determined by a differential scanning calorimeter (DSC). The results showed that the glass transition temperature ($T_g$) increased from 229 °C to 265 °C as the ratio PDMS was increased from 1:1 to 1:4. Moreover, thermogravimetric analyzer (TGA) was used to determine thermal stability of the blends. The TGA thermograms revealed that as the PDMS content increased, the char yield decreased from 23% (at a 1:1 ratio) to 12% (at a 1:4 ratio). The decomposition temperature at 10% weight loss ($T_{d10}$) increased from 362 °C to 392 °C as the PDMS was increased.

## 1. Introduction

Bio-based materials are currently gaining significant attention due to their potential to reduce environmental impacts compared to petroleum-based materials. These materials are biodegradable, safer for human health, and contribute to lower greenhouse gas emissions during production. They can enhance to response the requirements of diverse industries such as electronics, insulation, and sustainable technologies [1]. Previous studies have investigated bio-based benzoxazine materials derived from vanillin and furfurylamine, which are natural raw materials with dominant thermal stability and excellent mechanical properties. These characteristics make them a sustainable and renewable alternative to petroleum-based materials [2].

The thermal properties of bio-based materials are crucial for their practical applications, particularly in designing materials for high-temperature environments. Differential Scanning Calorimetry (DSC) is an essential technique for analyzing phase transitions such as the melting temperature and glass transition temperature ($T_g$), which directly impact the performance of materials. Meanwhile, Thermogravimetric Analysis (TGA) evaluates thermal stability and the decomposition behavior of materials when exposed to heat. These properties are important factors in designing and developing materials to effectively resist high-temperature applications [3,4].

In this study, bio-based benzoxazine (V-fa) was modified by mixing with polydimethylsiloxane (PDMS) to create a complex coupling structure using KH-550 as a crosslinking agent. A key highlight of this work is this study represents a novel approach, as there have been no prior reports on modifying benzoxazine structures using a crosslinking agent and the structural modification of V-fa to improve its flexibility, imparted by PDMS, while simultaneously enhancing its thermal properties, including high thermal stability and mechanical strength. These improved properties make the material suitable for applications requiring resistance to high temperatures. This research focuses on analyzing the thermal and mechanical properties of the developed materials, demonstrating their potential for heat-resistant applications.

## 2. Experimental

Vanillin-based benzoxazine (V-fa) was synthesized from vanillin, paraformaldehyde, and furfurylamine at a molar ratio of 1:2:1 using a solventless method [5]. The mixture of V-fa and PDMS were prepared at ratios of 1:1, 2:3, 1:2, and 1:4, with γ-aminopropyltriethoxysilane (KH-550) used as a crosslinking agent. These components were homogenously blended at 100°C for 10 minutes in an aluminium pan. Subsequently, the DBTDL catalyst (0.1 g) was added to the mixture, and the reaction was allowed to proceed at 100°C for 2 h., resulting in a pale-yellow solution. The mixture was degassed and then poured into mold. Curing was performed using a program of 120°C for 3 h., 150°C for 2 h., and 180°C for 2 h. in Figure 1 [6]. The chemical structure of the synthesized material was characterized using Fourier Transform Infrared Spectrometer (FT-IR). FT-IR analysis confirmed the formation of specific bonds between V-fa and PDMS. The curing conditions of the blends were analysed using a differential scanning calorimeter (Mettler Toledo, DSC822$^e$, Switzerland) at a heating rate of 20°C/min under a nitrogen atmosphere. The thermal stability of the samples was determined using a thermogravimetric analyzer (Mettler Toledo, TGA1 Model STAR$^e$ System, Switzerland). All samples were placed on crucible pans and heated from 25 to 800°C at a heating rate of 20°C/min. The decomposition temperatures and char yields at 800°C were reported.

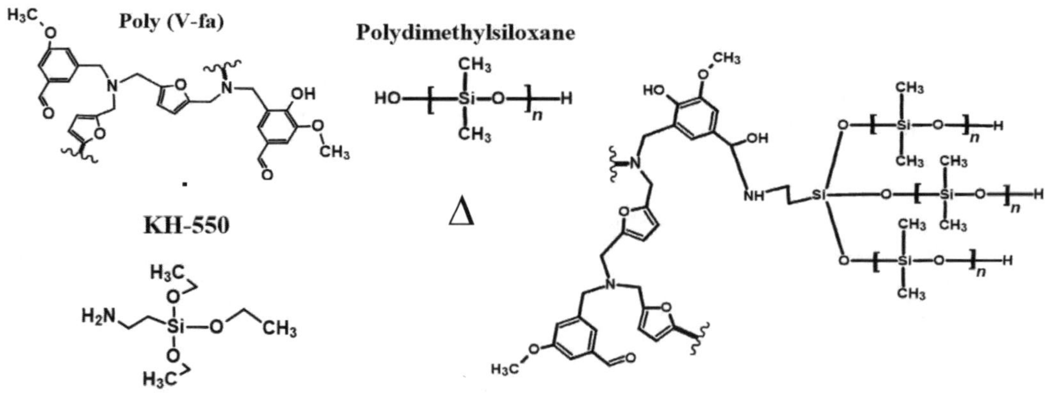


**Figure 1.** Crosslinking Reaction of Poly (V-fa) with KH-550 and Polydimethylsiloxane

### 3. Results and discussion

The FTIR spectra of the blends between V-fa and PDMS were showed in Figure 2. The reaction between two polymer was evidenced by several peak shifts. The peak at 960 $cm^{-1}$, corresponding to the benzoxazine ring, slightly decreased, suggesting ring-opening. Meanwhile, the peak at 1200 $cm^{-1}$ significantly increased, indicating the formation of the C-N-C bond in the Mannich bridge. Additionally, the disappearance of the peak at 3300 $cm^{-1}$ signifies that the OH group was consumed in the reaction. Furthermore, an increase in the peaks at 1054-1071 $cm^{-1}$, corresponding to the Si-O-Si structure from PDMS, and at 800-850 $cm^{-1}$, indicating Si-$CH_3$ from KH500, along with the pronounced rise in the 2940-3000 $cm^{-1}$ peak, showing $CH_3$ stretching, all confirm the bonding between V-fa and PDMS. The spectrum of the blend at 1:1 ratio showed the weak at 1254 $cm^{-1}$ reflects the formation of new bonds between the two substances, while the spectrum of the blend at poly (V-fa) showed decrease of C=O at 1680 $cm^{-1}$ indicates further reaction between them [6-9]. Together, these peak shifts confirm the success of the reaction between V-fa and PDMS.

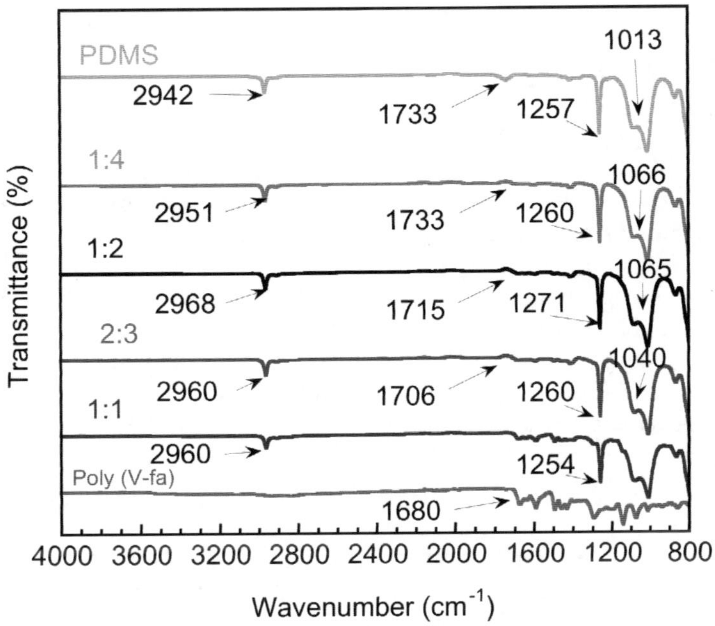


**Figure 2.** FTIR spectra for different V-fa:PDMS ratios

The thermal degradation analysis (TGA) of V-fa and its composites with PDMS at different ratios revealed that the thermal stability increased with higher PDMS content in Table 1. Poly (V-fa) started to degrade at $T_{d5}$ = 239°C and $T_{d10}$ = 314°C with a char yield of 56% at 800°C. For the V-fa:PDMS at 1:1 ratio, $T_{d5}$ and $T_{d10}$ increased to 361°C and 448°C, respectively, but the char yield decreased to 23%. In the V-fa:PDMS 1:2 ratio, thermal stability was like the at 1:1 ratio with $T_{d5}$ = 367°C, $T_{d10}$ = 449°C and a char yield of 24%. Finally, for the V-fa:PDMS at 1:4 ratio, $T_{d5}$ reached the highest value at 392°C, $T_{d10}$ = 463°C but the char yield was the lowest at 13%. In summary, adding more PDMS to V-fa improved its thermal stability, but the char yield decreased as the PDMS ratio increased [7,8].

**Table 1.** Degradation temperatures and char yield of V-fa blended with PDMS

| Weight ratio (V-fa:PDMS) | $T_{d5}$ (°C) | $T_{d10}$ (°C) | Char yield at 800 °C (%) |
|---|---|---|---|
| Poly (V-fa) | 239 | 314 | 56 |
| 1:1 | 361 | 448 | 23 |
| 1:2 | 367 | 449 | 24 |
| 1:4 | 392 | 463 | 13 |

The thermal properties were analyzed using a differential scanning calorimeter (DSC). Table 2 shows $T_g$ of the V-fa:PDMS increased as the weight ratio of PDMS was increased [3,9].

**Table 2.** Glass transition temperature of ration V-fa with PDMS

| Weight ratio (V-fa:PDMS) | Glass Transition Temperature ($T_g$, °C) |
|---|---|
| 1:1 | 229 |
| 2:3 | 237 |
| 1:2 | 249 |
| 1:4 | 265 |

## 4. Conclusions

In this research, the bio-based benzoxazine (V-fa) and polydimethylsiloxane (PDMS) were blended to create a complex coupling structure, and KH-550 was applied to enhance the structural strength. The successful reaction was confirmed through FTIR analysis, as evidenced by peak shifts indicating ring-opening, new bond formation, and the disappearance of functional groups. TGA revealed the increase of thermal stability with higher PDMS content, i.e., from $T_{d5}$ reaching 361°C ( at a 1:1 ratio) to 392°C (at a 1:4 ratio), though char yield decreased. DSC showed that the glass transition temperature gradually increased from 229 to 265°C as PDMS content. Therefore, the structural strength of the material was enhanced which, makes the material more suitable for applications requiring high durability.

**Acknowledgements**

This project is funded by National Research Council of Thailand (NRCT) and Chulalongkorn University (N42A660910). In addition, partial financial support provided from Department of Physical and Material Sciences, Faculty of Liberal Arts and Science, Kasetsart University was also acknowledged

**References**

[1] Ding H, Wang X, Song L and Hu Y 2022 *JRM* **10** 4.
[2] Prasomsin W, Parnklang T, Sapcharoenkun C, Tiptipakorn S and Rimdusit S 2019 *Nanomaterials.* **9** 881.
[3] Porras CL, Alcantar PC, Solis VE, Guerra EM, Pinon-Balderrama CI, Martinez IC and Saavedra-Leos MZ 2019 **12** 5.
[4] Farivar F, Yap PL, Karunagaran RU and Losic D 2021 *C. Journal of Carbon Research.* **7** 41.
[5] Ishida H 1996 *US patent* 5543516.
[6] Li Z, HuHu J, Ma L and Liu H 2019 *Applied Polymer.* 48621.
[7] Martincic M, Sandoval S, Oro-Sole J and Tobias-Rossell G 2024*Nanomerails* **14** 1754.
[8] Sha XL, Yang XZ, Fang S, Li Z, Li P and Liu Z 2024 *Reactive and Functional Polymer.* **196** 105832.
[9]Ardhyananta H, Haniff Wahid M, Sasaki M, Agag T, Kawauchi T, Ismail H, and Takeichi T 2008 *polymer* **49** 4589.

# Effects of molecular weight of polycaprolactone on the thermal properties of V-fa/PCL blends added with cross-linking agent

**N Chaipakdee[1], K Hemvichian[2], P Lertsarawut[2], S Rimdusit[3], and S Tiptipakorn[1,*]**

[1] Department of Physical and Material Sciences, Faculty of Liberal Arts and Science, Kasetsart University, Nakhon Pathom, 73140, Thailand.
[2] Thailand Institute of Nuclear Technology (Public Organization), Ongkarak District, Nakornnayok 26120, Thailand.
[3] Center of Excellence in Polymeric Materials for Medical Practice Devices, Department of Chemical Engineering, Faculty of Engineering, Chulalongkorn University, Bangkok, 10330, Thailand.

*Corresponding author: faassntk@ku.ac.th

**Abstract.** In this research, the blends between bio-based polybenzoxazine (V-fa) and polycaprolactone (PCL) with different molecular weights ($M_n$) of 14,000 and 80,000 Da were prepared at the PCL contents varied at 80, 90, and 95 wt%. γ-Aminopropyltriethoxysilane (KH-550) was used as cross-linking agent. The chemical structures of the polymer blends were characterized by Fourier Transform Infrared Spectrometer (FT-IR). The spectra revealed that hydrogen bonds between two polymers were observed. Moreover, the V-fa and PCL were cross-linked when KH-550 was applied. Thermal properties were determined by thermogravimetric analyzer (TGA). The results of the blends with KH-550 describes the temperature at maximum decomposition rate ($T_{d,\,max}$) and char yield. In addition, at the same molecular weight, when the contents of PCL increased, $T_{d,\,max}$ was slightly increased from 426 to 429 °C, but char yield was decreased. However, these values had no significant difference when no KH-550 was used. Moreover, when bending the V-fa/PCL at a $M_n$ of 80,000 Da, the sample was flexible and could be bent.

## 1. Introduction

Polybenzoxazine (PBZ) is one of high-performance synthetic polymers with excellent mechanical, electrical and thermal properties and good processability. PBZ is a new type of polymeric phenolic resin with remarkable properties, including low viscosity, ease of synthesis, and flame resistance [1]. PBZ, novel type of thermosetting, is prepared through ring-opening polymerization of benzoxazine monomers using phenol derivatives, aldehydes, and amine groups as precursors. However, it also has limitations, such as brittleness, rigidity, and non-biodegradability [2]. To address these drawbacks, in this study, bio-based polybenzoxazine (V-fa) was modified with polycaprolactone (PCL) of number average molecular weight ($M_n$) of 14,000 (14k) and 80,000 (80k) Dalton with various PCL contents. PCL acts as an additive to enhance the processability and other properties, such as impact resistance, and flexibility. Moreover, due to its excellent biocompatibility, PCL could apply in medical fields [3].

From the literature [4], the polybenzoxazine and polycaprolactone system without cross-linking agent, at number average molecular weight ($M_n$) of PCL at 10,000 Da was studied. However, the effect of the molecular weight of PCL in the presence of cross-linking agent have not been studied yet. Therefore, in this study, bio-based polybenzoxazine (V-fa) was developed by blending with polycaprolactone (PCL) and using KH-550 as a cross-linking agent. The chemical interactions between two polymers were investigated via Fourier Transform Infrared Spectroscopy (FT-IR). The thermal properties were determined with differential scanning calorimetry (DSC), thermogravimetric analysis (TGA), and bending tests.

## 2. Experimental

Bio-based benzoxazine (V-fa) monomer was synthesized from vanillin, furfurylamine, and paraformaldehyde at a molar ratio of 1:1:2. The mixture was then heated at 105 °C for 1 h. The V-fa monomer and PCL at different number average molecular weights ($M_n$) of 14,000, and 80,000 Da were blended at the PCL contents varied at 80, 90, and 95 wt.% in a vial. KH-550 at 5 wt.% was used as cross-linking agent, after homogeneous mixture was obtained were poured into a mold and a step-cured process at temperature of 150 °C, 160 °C (for each 1 h), and 170 °C, 180 °C (for each 2 h), respectively [5]. After cooling down to room temperature, V-fa/PCL blends were obtained and then characterized. The Fourier Transform Infrared Spectrometer (FT-IR) in Attenuated Total Reflectance (ATR) mode. The thermal properties were determined by a thermogravimetric analyzer (Mettler Toledo, TGA1 STAR$^e$ System, Switzerland), a differential scanning calorimeter (Mettler Toledo, DSC882$^e$, Switzerland). After that, bending test was conducted. The code names to present the sample with/without the crosslink agent are presented in the Table 1.

**Table 1.** Sample code

| Sample code | $M_n$ of PCL (Da) | V-fa (wt.%) | PCL (wt.%) | Cross-link agent (wt.%) |
|---|---|---|---|---|
| 14kP90wo | 14,000 | 10 | 90 | 0 |
| 14kP80c | 14,000 | 15 | 80 | 5 |
| 14kP85c | 14,000 | 10 | 85 | 5 |
| 14kP90c | 14,000 | 5 | 90 | 5 |
| 80kP80c | 80,000 | 15 | 80 | 5 |
| 80kP85c | 80,000 | 10 | 85 | 5 |
| 80kP90c | 80,000 | 5 | 90 | 5 |

## 3. Results and discussion

The molecular structure of V-fa/PCL blends was investigated by using ATR FT-IR spectroscopy. The spectra are shown in Figure 1(a). The spectrum of the 14kP90c showed peaks at 1726 $cm^{-1}$ of C=O stretching and 1167 $cm^{-1}$ of C-O-C ester. The spectrum of the 14kP90wo showed peaks at 1721 and 1176 $cm^{-1}$. The peaks of carbonyl and ester groups were shifted, which revealed the presence of hydrogen bonds between two polymers. Moreover, the V-fa and PCL were cross-linked when KH-550 was applied. Other spectra could be observed at 1682 and 1674 $cm^{-1}$, corresponding to C=O stretching of poly(V-fa) and V-fa monomer, respectively [6]. It is proposed that this is the formation of hydrogen bonds between V-fa/PCL and V-fa/PCL with KH-550 as shown in Figure 2(b).

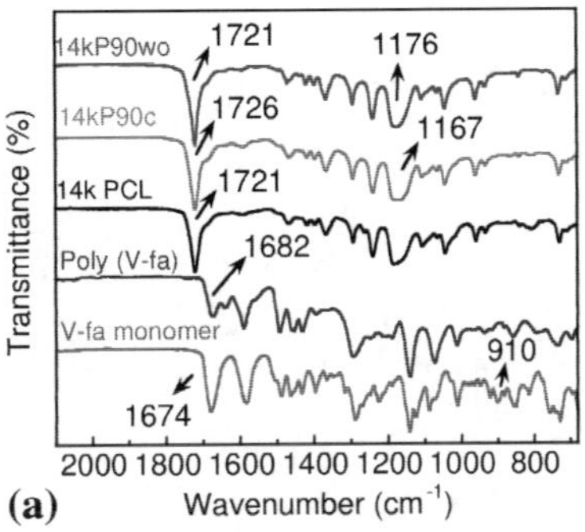

(a)

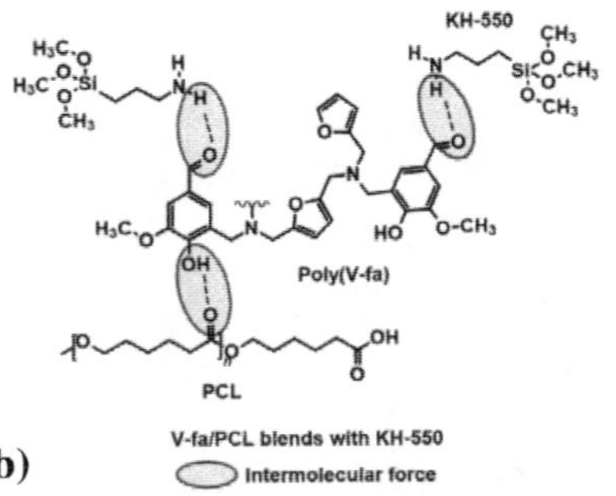

(b)

**Figure 1**. ATR FT-IR spectra of V-fa/PCL blends (a) and proposed chemical structure of V-fa/PCL blends with KH-550 (b).

The results from DSC analysis revealed that the melting temperature ($T_m$) values of the 80k PCL blends were slightly higher (i.e. 1-2 °C higher) than those of the 14k PCL blends with values in the range of 62 to 68 ˚C. This could be attributed to higher hydrogen bonding of 80 k PCL. The phenomenon corresponded to the similar system of PBA-a/80kPCL [7]. For the heat of fusion ($H_f$), at the same molecular weight, increasing the amount of PCL led to a higher $H_f$ due to more requirement of heat of melting However, when the molecular weight increased, at the same PCL content, the heat of fusion ($\Delta H_f$) values decreased because the crystallinity of PCL with a molecular weight of 14k was higher than that of 80k are presented in the Table 2. This is consistent to our previous work [7]. That reported lower molecular weight PCL formed larger spherulite size and higher crystallinity, leading to more heat absorption.

**Table 2.** DSC results of the V-fa/PCL at PCL contents of 80 to 95 wt.% at $M_n$ 14k and 80k.

| Sample code | $T_m$ (˚C) | Heat of fusion (J/g) |
|---|---|---|
| 14kP80c | 64 | 116.2 |
| 14kP85c | 62 | 167.5 |
| 14kP90c | 66 | 188.5 |
| 80kP80c | 66 | 99.4 |
| 80kP85c | 68 | 147.6 |
| 80kP90c | 67 | 165.9 |

The thermal stability was determined by TGA; the results showed that poly(V-fa) demonstrated a $T_{d,max}$ of 408°C and a char yield of 56%. Figure 2 showed the comparison of these values of the 14kP80c, 14kP85c, and 14kP90c. It was found that the $T_{d,max}$ values were 426, 427, and 429 °C, respectively. The char yield values were 10, 6, and 2 %, respectively. It was observed that the increase of the PCL content resulted in the increase of $T_{d,max}$ and the decrease of the char yield. This is attributed to the increase of thermal stability from stronger intramolecular force, leading to slower decomposition and a reduced amount of residue.

Conversely, with different PCL molecular weights, there was no significant difference in $T_{d,max}$ and char yield of the blends at the same PCL content. This phenomenon aligns with the report of our previous work [7], studying the PBA-a and PCL systems with various molecular weights. They concluded that molecular weight has no effect on the thermal stability of the blends.

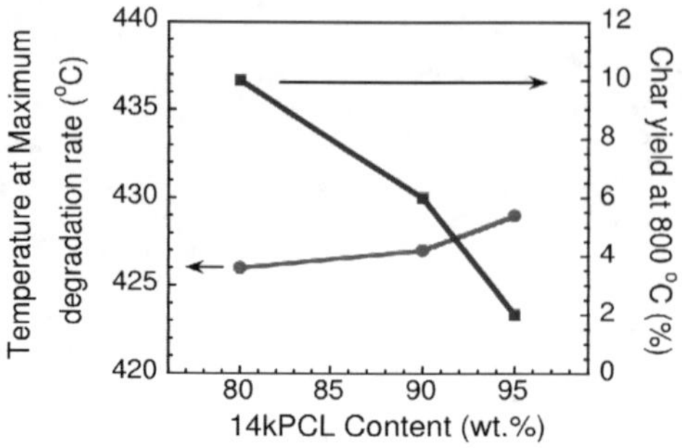


**Figure 2**. TGA results of V-fa/PCL blends added with KH-550.

The bendability tests of the 14kP95c and 80kP95c was conducted. The results revealed that the blend at the molecular weight of 14k Da was brittle and could not be bendable, On the other hand, the blend at the molecular weight of 80k Da was flexible and could be bendable, as shown in Figure 3.

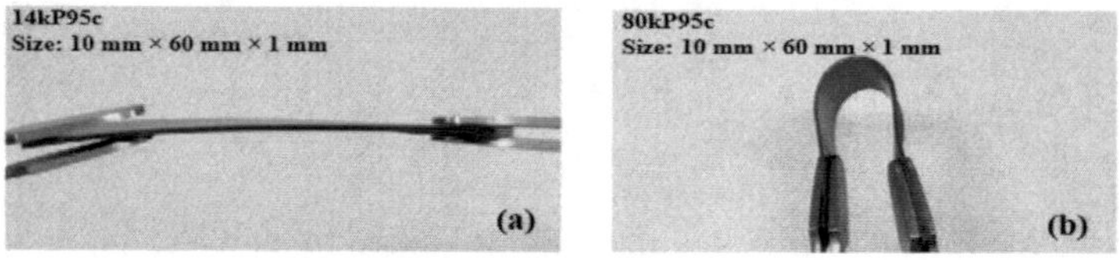


**Figure 3**. Photographs of V-fa/PCL blends of 14kP95c (a), and 80kP95c (b).

## 4. Conclusions

In this research, bio-based polybenzoxazine (V-fa) was synthesized and mixed with polycaprolactone (PCL) at different molecular weights of 14k and 80k Da. KH-550 was used as a cross-linking agent. The brittleness of polybenzoxazine was improved significantly with such modifications. The formation of hydrogen bonds could be found, which could improve thermal properties and flexibility. From the results of thermal properties, there was no significant difference for the blends at different values of $M_n$; however, the thermal properties depended on PCL contents. Moreover, it was found that only the blends at 80k Da were flexible and could be bendable. The blends have potentials to be applied as sensor, flexible material, insulation and microelectronic.

## Acknowledgements

This project is funded by National Research Council of Thailand (NRCT) and Chulalongkorn University (N42A660910). In addition, partial financial support provided from Departments of Physical and Material Sciences, Faculty of Library Arts and Science, Kasetsart University was also acknowledged.

## References

[1] Ding H, Wang X, Song L and Hu Y 2022 *Journal of Renewable Materials*. **10** 4.
[2] Amornkitbamrung L, Leungpuangkaew S, Panklang T, Jubsilp C, Ekgasit S, Um S H and Rimdusita S 2022 *Journal of Science: Advanced Materials and Devices*. 100446.
[3] McKeen L 2021 *The Effect of Long-Term Thermal Exposure on Plastics and Elastomers (Second Edition)*. **313** 332.
[4] Huang J H and Yang S J 2005 *Polymer*. **46** 8068
[5] Hombunma P, Parnklang T, Mora P, Jubsilp C and Rimdusit S 2019 *Smart Materials and Structure*. 108447.
[6] Amornkitbamrung L, Srisaard S, Jubsilp C, Bielawskia W B, Um S H and Rimdusita S 2020 *Polymer*. **209** 122986.
[7] Tiptipakorn S, Keungputpong N, Phothiphiphit S and Rimdusit S 2015 *Journal of Applied Polymer science*. **22** 132.

IOP Conf. Series: Materials Science and Engineering **1325** (2025) 012004 doi:10.1088/1757-899X/1325/1/012004

# Atomistic Simulations of 5CB Liquid Crystal on *Trans*- and *Cis*-Azo Interfaces

**T. Fudulwatjananon[1,2], N. Nisoh[1,2], P. Nalakarn[2,3] and J. Wong-Ekkabut[1,2,*]**

[1] Department of Physics, Faculty of Science, Kasetsart University, Bangkok 10900, Thailand.
[2] Computational Biomodelling Laboratory for Agricultural Science and Technology (CBLAST), Faculty of Science, Kasetsart University, Bangkok 10900, Thailand.
[3] Department of Physics, Faculty of Science and Technology, Thammasat University, Pathumthani 12120, Thailand

*Corresponding author: jirasak.w@ku.th

**Abstract.** In this study, we examine the surface anchoring behaviour of 4-Pentyl-4'-Cyanobiphenyl (5CB) liquid crystals on *trans*- and *cis*-azo interfaces using united-atom molecular dynamics simulations. Our findings suggest homeotropic anchoring at the *trans*-azo surface and random planar anchoring at the *cis*-azo surface. This study provides a better understanding of how *trans*- and *cis*-azo interfaces affect the orientation of the LC molecules.

## 1. Introduction

The orientation of liquid crystal is a defining characteristic of liquid crystals that controls their optical properties. The orientations of liquid crystals are influenced by their interaction with surfaces. Interestingly, azo molecule and its derivatives can reversibly change their formation from stable *trans*- to *cis*- when exposed to UV light, a process known as photoisomerization. The photoisomerization reaction of azobenzene and its derivatives has been shown to reversibly switch the alignment of nematic liquid crystals, controlling liquid crystal's orientation [1, 2]. However, the interaction between azo molecules and liquid crystals at the molecular level is still unclear.

In recent years, molecular dynamics (MD) simulations have been used to study the behaviour and dynamics of liquid crystal molecules [3-5]. This technique can provide insights into the surface anchoring mechanisms of liquid crystals by examining the systems at the nanoscale, which is difficult to observe by experiment. In this research, we study the effects of *trans*- and *cis*-azo (Azo5, a derivative of azobenzene) surfaces on 5CB liquid crystal. We examine the orientational order and dynamic properties to obtain a deeper understanding of the surface anchoring behaviour of liquid crystal molecules.

## 2. Model and simulation method

To investigate the anchoring behaviour of *trans*- and *cis*-azo surfaces, molecular dynamics simulations were prepared for each surface. Both systems consist of 1000 5CB LC molecules confined between 36 molecules of azo surface and vacuum surface. The positions of H1 (*cis*-azo) and H16 (*trans*-azo) atoms are restrained using a spring constant of 1000 $kJ \cdot mol^{-1}$. The angles of H1-C8 (*cis*-azo) and H16-C10

IOP Conf. Series: Materials Science and Engineering **1325** (2025) 012004 doi:10.1088/1757-899X/1325/1/012004

(*trans*-azo) are restrained to align parallel to the z-axis using the same spring constant. The initial configurations of the systems are shown in Fig 1.

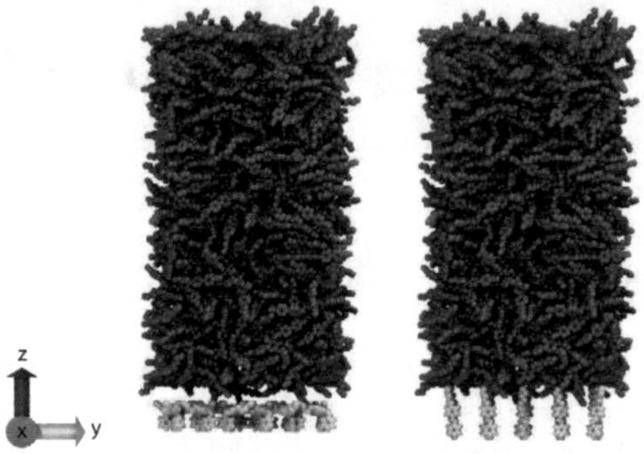


**Figure 1.** The initial configurations of *cis*-azo systems (left) and *trans*-azo systems (right) where 5CB are confined between the azo surface at the bottom and the vacuum region at the top. Green represents azo molecules, and grey represents 5CB molecules

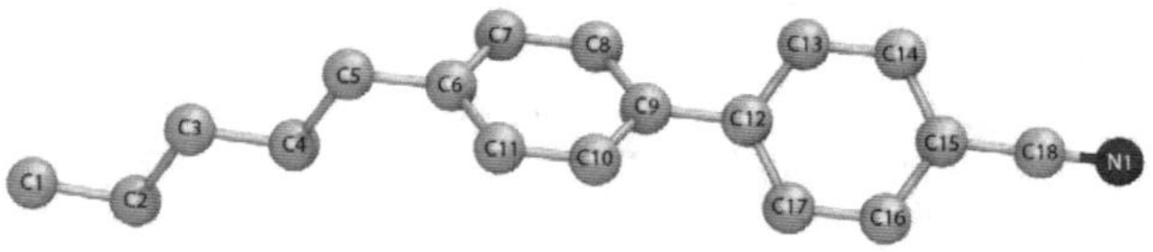


**Figure 2.** The united-atom model of 5CB molecules, where nitrogen atoms are represented in blue, and carbon atoms are represented in cyan.

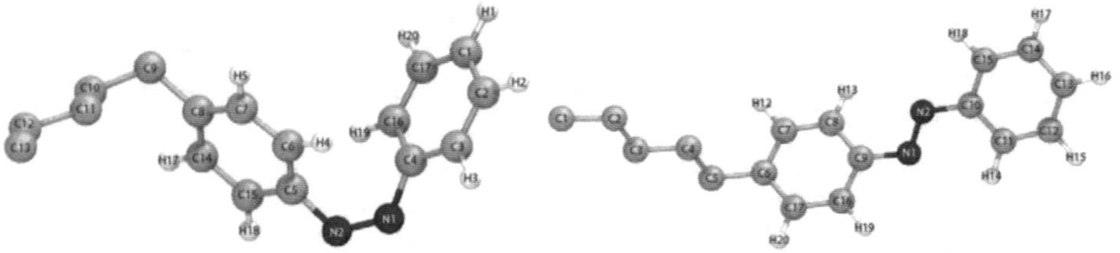

**Figure 3.** The united-atom model of *cis*-azo molecules (left) and *trans*-azo molecules (right), where nitrogen atoms are represented in blue, carbon atoms are represented in cyan, and hydrogen atoms are represented in white.

The united-atom force field for 5CB molecules was produced using parameters from Zhang et al. [6], and force fields for *trans*- and *cis*-azo molecules were calculated from the Automated Topology Builder (ATB) and Repository [7]. The united-atom molecular models of 5CB, *cis*-azo and *trans*-azo molecules are shown in Fig 2 and Fig 3, respectively.

Simulations were performed under the NVT ensemble at 300 K using GROMACS 5.1.2 with a 1 ps time step. The velocity rescale thermostat [9] was applied to maintain a constant temperature, with a coupling time of 0.1 ps. The particle-mesh Ewald (PME) method [10] was used to calculate the long-range electrostatic interactions. The simulation boxes have the surface dimensions of 60 Å × 60 Å with a thickness of approximately 120 Å. The simulations were run for 400 ns.

## 3. Results and discussion

The order parameter **S** measures how well the liquid crystal molecules align with a preferred direction, and it can be calculated from the largest eigenvalue of the ordering matrix **Q** which is expressed as [11]

$$\mathbf{Q} = \frac{3}{2}\left\langle \mathbf{a} \otimes \mathbf{a} - \frac{1}{3}\mathbf{I} \right\rangle \quad (1)$$

where **a** is the long molecular axis unit vector of 5CB molecule, which we defined as unit vector from C6 pointing to N1. **I** is the identity matrix, and $\langle ... \rangle$ is averaging on all liquid crystal molecules.

The order parameter as a function of time for the *cis*- and *trans*-azo systems is shown in Fig 4. Both systems start in an isotropic phase ($\mathbf{S} \approx 0$) and transition into a nematic phase. Over time, the order parameter S stabilizes, with the *cis*-azo system converging to approximately 0.28±0.07 and the *trans*-azo system converging to approximately 0.45±0.10, based on the average over the last 100 ns. The results indicate that the *cis*-azo surface results in a less ordered compared to the *trans*-azo surface.

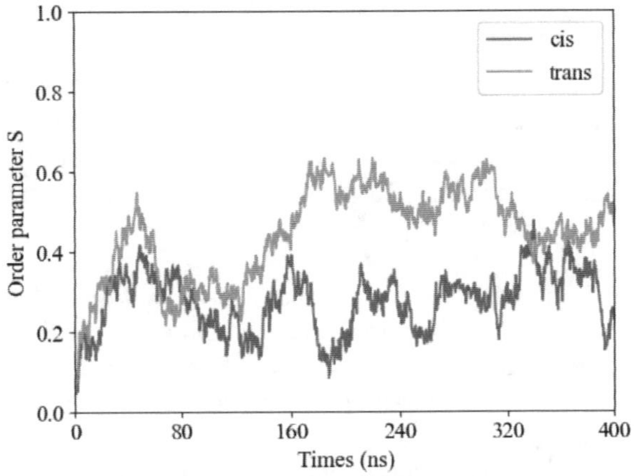


**Figure 4.** The orientational order parameter **S** in function of time for *cis*- and *trans*-azo systems.

The final configurations of the systems are shown in Fig 5. The green molecules represent azo molecules. The colours of the 5CB indicate their angle formed by 5CB molecular axis (C6 to N1) relative to the z-axis: blue (0°–60°), grey (60°–120°), and red (120°–180°). In both systems, 5CB exhibits homeotropic anchoring at the vacuum interface. At the azo surface, 5CB interpenetrates, but only 5CB on the *trans*-azo surface shows clear homeotropic anchoring.

**Figure 5.** The final configurations of *cis*-azo system (left) and *trans*-azo system (right). Green represents azo molecules. The colours of the 5CB indicate their angle to z-axis: blue (0°–60°), grey (60°–120°), and red (120°–180°).

The dynamic properties were analysed by using the centre-of-mass of 5CB molecules to calculate the mean squared displacement (MSD) and diffusion coefficient for 10 ns. The results of the mean squared displacement are shown in Figure 6.

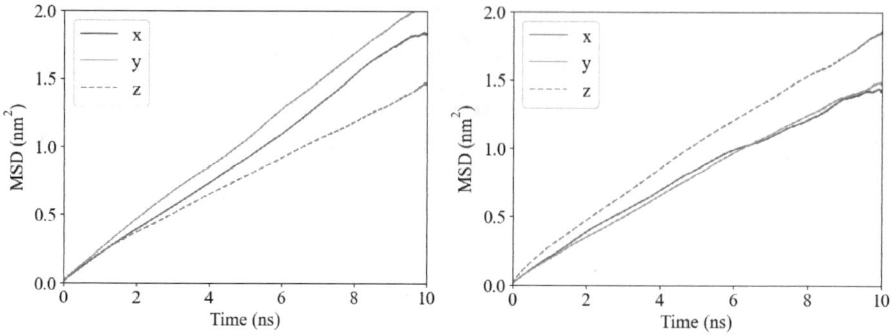


**Figure 6.** Mean square displacements of 5CB for *cis*-azo system (left) and *trans*-azo system (right).

For the *trans*-azo system, the diffusion coefficients of 5CB are $8.8\times10^{-11}$ m$^2$/s and $7.3\times10^{-11}$ m$^2$/s for parallel and lateral to the z-axis, respectively. For the *cis*-azo system, the values are $6.8\times10^{-11}$ m$^2$/s and $9.8\times10^{-11}$ m$^2$/s for parallel and lateral to the z-axis respectively. The higher parallel diffusion in the *trans*-azo system indicates homeotropic anchoring. Conversely, the higher lateral diffusion in the *cis*-azo system and nearly identical diffusion in the x and y directions, indicates random planar anchoring.

## 4. Conclusions

In this research, we have performed the molecular dynamics simulation and exhibited that the *trans*- and *cis*-azo surfaces crucially affect the orientation and dynamics of 5CB liquid crystals. This work illustrates that the *trans*-azo surface promotes strong homeotropic anchoring, leading to increased molecular order and decreased lateral diffusion. However, the *cis*-azo surface promotes random planar anchoring, leading to decreased molecular order and increased lateral diffusion. These findings show the role of surface structure in controlling liquid crystal orientation and dynamics. Also offering insights for developing and designing advanced liquid crystal devices.

## Acknowledgement

This work was supported by Development and Promotion of Science and Technology Talents Project (DPST), the National Science, Research and Innovation Fund (NSRF) via the Program Management Unit for Human Resources & Institutional Development, Research and Innovation [grant number B13F660122], and Faculty of Science Kasetsart University

## References

[1] Ichimura K, Suzuki Y, Seki T, Hosoki A and Aoki K 1988 *Langmuir* **4** 1214
[2] Eremin A, Nádasi H, Hirankittiwong P, Kiang-Ia J, Chattham N, Haba O, Yonetake K and Takezoe H 2018 *Liq. Cryst.* **45** 2121
[3] Sadati M, Ramezani-Dakhel H, Bu W, Sevgen E, Liang Z, Erol C, Rahimi M, Taheri Qazvini N, Lin B and Abbott N L 2017 J. Am. Chem. Soc. **139** 3841
[4] Ramezani-Dakhel H, Sadati M, Rahimi M, Ramirez-Hernandez A, Roux B and de Pablo J J 2017 *J. Chem. Theory Comput.* **13** 237
[5] Chen H, Xu C, Xiao G, Chen Z, Yi M and Zhang J 2020 *Appl. Surf. Sci.* **508** 145284
[6] Zhang J, Su J and Guo H 2011 *J. Phys. Chem. B.* **115** 2214
[7] Malde A K, Zuo L, Breeze M, Stroet M, Poger D, Nair P C, Oostenbrink C and Mark A E 2011 J. Chem. Theory Comput. **7** 4026
[8] Abraham M J, Murtola T, Schulz R, Páll S, Smith J C, Hess B and Lindahl E 2015 *SoftwareX* **1** 19
[9] Bussi G, Donadio D and Parrinello M 2007 *J. Chem. Phys.* **126**
[10] Darden T, York D and Pedersen L 1993 *J. Chem. Phys.* **98** 10089
[11] Luckhurst G R and Veracini C A 2012 *The molecular dynamics of liquid crystals vol 431: Springer Science & Business Media)*

# Smart sensing film developed by polyvinyl alcohol/carboxymethyl cellulose doped with anthocyanin extract as a natural pH indicator

**P Rieagnam[1], P Jaisabai[1], K Wongkhan[1] and R Jitchati[1]***

[1] Department of Chemistry, Faculty of Science, Ubon Ratchathani University, Ubon Ratchathani 34190, Thailand

*Corresponding author: rukkiat.j@ubu.ac.th

**Abstract.** This research focused on developing a pH-responsive film based on a poly(vinyl alcohol)/carboxymethyl cellulose (PVA/CMC) composite combined with anthocyanin, a natural pH indicator extracted from butterfly pea flower (ABA). Coconut husks were treated and bleached with NaOH and $H_2O_2$ before undergoing carboxymethylation. The bleached coconut husks were converted into CMC through alkalization, followed by etherification with NaOH and $ClCH_2COOH$ in an alcoholic medium. The CMC/PVA film, incorporating ABA as a natural pH indicator, was prepared via casting with varying anthocyanin contents. The PVA-to-CMC ratio and ABA content significantly influenced the mechanical properties and pH sensitivity of the ABA-PVA/CMC films. Notably, films with a PVA/CMC ratio of 4:1 exhibited the highest tensile strength and uniform dispersion. Additionally, ABA-PVA/CMC films displayed strong pH sensitivity, changing color distinctly at different pH levels.

## 1. Introduction

Recently, there has been an increasing emphasis on ensuring food safety and maintaining high-quality standards has driven the advancement of intelligent packaging systems capable of providing real-time monitoring and feedback for the transportation, consumption, and storage of food [1-2]. One notable approach in smart packaging is the use of pH-responsive indicators, which allow consumers to assess the product quality without direct contact by signaling changes in the product inside the package, offering a simple and effective means of monitoring freshness [5]. During the storage of high-protein foods such as meat, dairy, and seafood, a significant amount of evaporative alkaline compounds, including substances like dimethylamine, trimethylamine, and ammonia, are produced, these compounds can create an alkaline environment in the headspace of the packaging [6-7]. Generally, pH colorimetric indicators are composed of two key components: polymeric carrier matrix supports and pH-sensitive dyes [3-4].

Historically, synthetic compounds have been used as pH colorimetric indicators. However, these substances are not designed to be consumable, and their direct contact with food could present potential health risks. Due to concerns over their toxicity, their application in food quality monitoring is limited, as they may pose a danger to consumers if ingested. Recently, natural pH dyes have gained more attention since they are non-toxic and environmentally friendly properties, in contrast to chemo-responsive dyes [8]. Butterfly pea flowers are an herb known for their high anthocyanin content.

Anthocyanin from butterfly pea flowers shows remarkable properties due to its sensitivity to both acidic and alkaline conditions. Butterfly pea flowers (*Clitoria ternatea*) are rich in various bioactive compounds, notably anthocyanins, which are responsible for their blue color [9-10]. These compounds are powerful antioxidants and have multiple potential health benefits. Hence, in this study, anthocyanins from butterfly pea flowers (ABA) were utilized as a pH-sensitive dye to create a smart colorimetric film.

This study explores the development of a pH-responsive film by combining polyvinyl alcohol (PVA) and carboxymethyl cellulose (CMC) with anthocyanin, a natural pigment extracted from butterfly pea flowers, known for its vivid color changes in response to pH variation, as presented in Figure 1. Anthocyanin, serving as a natural pH indicator, imparts distinct colorimetric responses to the composite films, offering a sustainable and low-cost alternative to synthetic indicators. Furthermore, coconut husk, an abundant waste material, was utilized to produce CMC through a series of chemical treatments, adding an additional layer of sustainability to the composite film development. The primary aim of this work is to examine the effects of varying PVA-to-CMC ratios and anthocyanin concentrations on the mechanical properties and pH sensitivity of the resulting films. By systematically adjusting these parameters, we aim to identify the optimal composition for producing films that combine both robust mechanical strength and effective pH sensitivity. This work not only highlights the potential of PVA/CMC/anthocyanin composites in creating sustainable, responsive materials but also provides a deeper understanding of the interplay between composition and performance in bio-composite film systems.

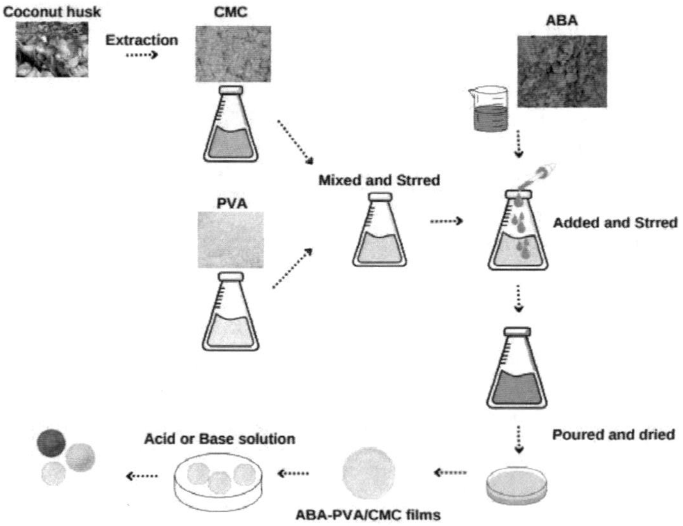


**Figure 1**. Schematic diagram of preparation of ABA-PVA/CMC film for response to pH change

## 2. Experimental

### 2.1 Materials

Coconut husk was obtained from Charoensri Market, Ubon Ratchathani, Thailand. The chemicals used sodium hydroxide (NaOH), hydrogen peroxide ($H_2O_2$), chloroacetic acid ($ClCH_2COOH$), hydrochloric acid (HCl), nitric acid ($HNO_3$), ammonia ($NH_4OH$), and acetic acid ($CH_3COOH$) were purchased from Italmar (Thailand) Co. Ltd., Bangkok, Thailand. Polyvinyl alcohol (PVA) was purchased from Unilab Pharmaceuticals Co. Ltd., Bangkok, Thailand. Methanol and acetone, the solvents, were obtained from Chemical and Material Co. Ltd., Bangkok, Thailand.

### 2.2 Preparation of cellulose from coconut husk
Coconut husk was washed, cut into small pieces, and then dried at 80 °C. The dried husk was mashed and sieved through a 40-mesh sieve. A total of 90 g of coconut husk powder was then mixed with 1000 mL of a 5% w/v NaOH solution and heated at 60 °C for 2 hours. Once the reaction was complete, the obtained brown slurry was filtered and washed with deionized water until the pH reached 7. The residue was dried in an oven at 60 °C to obtain cellulose. The cellulose was then bleached using a 20% w/v NaOH solution and a 30% v/v $H_2O_2$ solution (the mass ratio of NaOH to $H_2O_2$ was 1:5). This mixture solution was heated at 80 °C for 2 hours. The white slurry obtained was filtered and washed with deionized water until the pH reached 7. The cellulose powders were dried and stored for later use. Finally, the dried samples were characterized using FTIR.

### 2.3 Preparation of carboxymethyl cellulose (CMC)
A 4 g of cellulose powder was mixed with 20 mL of a 2% w/v NaOH solution in ethanol (40 mL). The resulting mixture was then added to a chloroacetic acid solution (6 g of chloroacetic acid in 70 mL of ethanol). The mixture solution was heated at 80 °C for 6 hours. Once the reaction was complete, the brown slurry was filtered and washed with deionized water until the pH reached 7. The residue was dried in an oven at 60 °C to obtain CMC. Finally, the dried CMC samples were characterized using FTIR.

### 2.4 Extraction of anthocyanin from butterfly pea flower (ABA)
A 20 g of butterfly pea flowers was mixed with 2000 mL of deionized water in a boiling pot and extracted at 50 °C for 60 minutes. The resulting supernatant was then filtered and stored for later use at 4 °C. The collected supernatant was freeze-dried to obtain dried anthocyanin from butterfly pea flowers (ABA). Finally, the dried ABA samples were characterized using FTIR.

### 2.5 Preparation of ABA-PVA/CMC smart sensing colorimetric films
The colorimetric films based on ABA-PVA/CMC were fabricated using a casting technique, as described in previous studies [11]. Initially, 8 g of polyvinyl alcohol (PVA) was dissolved in 100 mL of deionized water and heated to 90 °C while stirring until a clear, transparent solution was formed. Meanwhile, 2.0 g of carboxymethyl cellulose (CMC) was dissolved in 100 mL of deionized water and stirred at 75 °C until a transparent solution was achieved. The PVA and CMC solutions were then combined in a 1:1 (v/v) ratio at 70 °C, and the mixture was stirred until it became homogeneous. Next, varying contents of ABA powder (0, 0.2, 0.4, and 0.6 g) were dissolved in a small volume of deionized water and gradually added dropwise to the PVA/CMC solution. This resulted in final ABA concentrations of 0, 20, 40, and 60 mg/100 mL, respectively. The solution was stirred for 30 minutes to ensure complete dispersion of the ABA powder throughout the mixture. Finally, 10 mL of the prepared solution was poured into a petri dish and dried in an oven at 40 °C for 24 hours, allowing the films to form. The resulting films were designated as PVA/CMC, 0.2-ABA-PVA/CMC, 0.4-ABA-PVA/CMC, and 0.6-ABA-PVA/CMC, corresponding to ABA powder contents of 0, 0.2, 0.4, and 0.6 g, respectively.

### 2.6 Physicochemical properties of ABA-PVA/CMC films
FTIR (Spectrum RX I) in Attenuated Total Reflectance (ATR) mode with a diamond crystal was used to determine the chemical structure of the samples. The mechanical properties of the smart sensing colorimetric films were evaluated using a Dynisco universal testing machine (CG Engineering Co., Ltd., Thailand) with a head speed of 500 mm/min and a 1 kN load cell. The tensile strength test of the film compounds was carried out according to ASTM D412 standards. At least five samples of each film material were analyzed, and the results are presented as the average of the samples tested. The morphological characteristics of the smart sensing colorimetric films were examined using a scanning electron microscope (SEM, JSM-6010LV, Jeol Ltd., Tokyo, Japan) at an acceleration voltage of 15 kV. The fractured surfaces of the films from mechanical testing were coated with platinum before imaging.

IOP Conf. Series: Materials Science and Engineering **1325** (2025) 012005 doi:10.1088/1757-899X/1325/1/012005

### 2.7 The ABA-PVA/CMC film's response to pH changes
The film samples were cut into small square pieces (2 x 2 cm) and then immersed in a 3 mol/L $NH_4OH$ and 3 mol/L $CH_3COOH$ solution. Once fully immersed, the pieces were immediately removed and placed on filter paper. After 2 minutes, the color change were observed and photographed. The gas sensors should quickly respond to analytes.

## 3. Results and discussion
### 3.1 Physical appearance of cellulose, CMC, and ABA
The physical characteristics of cellulose, CMC, and ABA Results and discussion derived from coconut husk are shown in Figure 2. Dark brown and white powders were obtained after extracting cellulose from the coconut husk, as presented in Figure 2 (a) and (b), respectively. The yield of the obtained cellulose was 42%, which is similar to the result reported by Abdul et. al. [14]. The modification of cellulose to carboxymethyl cellulose (CMC) resulted in yellow powder, as shown in Figure 2 (c). The yield of the obtained CMC was 75%. The physical characteristics of anthocyanin powders extracted from butterfly pea flowers are shown in Figure 2 (d). The obtained anthocyanin (ABA) exhibited a purple color, with a yield of approximately 63%.

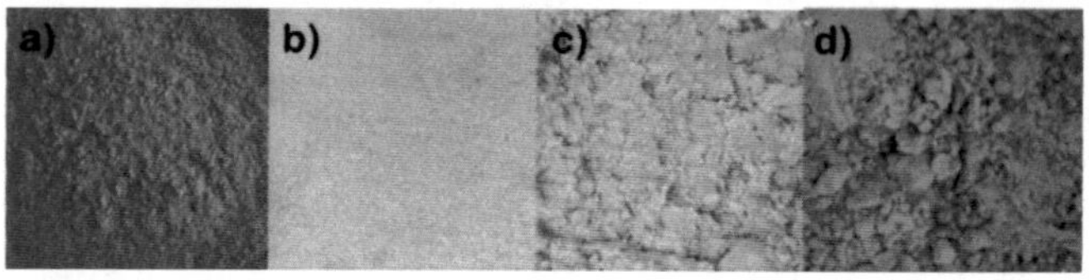


**Figure 2**. Photographs of (a) cellulose after the alkali treatment, (b) the cellulose after the bleaching, (c) carboxymethyl cellulose, and (d) anthocyanin extracted from butterfly pea flower.

### 3.2 Structure characterization
The functional groups were analyzed using FT-IR measurements. The spectra of coconut husks, isolated cellulose, and carboxymethyl cellulose (CMC), as presented in Figure 2. The spectrum of coconut husk displayed a broad absorption peak at 3175–3490 $cm^{-1}$ and a smaller peak at 2860–2938 $cm^{-1}$, which are characteristic of -OH stretching and C-H stretching, respectively. The absorption peaks at 1240 $cm^{-1}$ and 1023 $cm^{-1}$ are attributed to the C-O-C stretching and C-O stretching vibrations, respectively [12-13]. The peak at 1640 $cm^{-1}$ corresponds to the vibrations of water molecules absorbed in the cellulose [14]. We observed that lignin and hemicellulose were removed from the coconut husk, as indicated by the literature values at 1734, 1420, and 1512 $cm^{-1}$ [15-16]. Additionally, carboxymethyl cellulose (CMC), shown in Figure 3c, exhibited a new peak at 1731 $cm^{-1}$, characteristic of the ester carbonyl ($COO^-$) group, indicating that the hydroxyl groups of cellulose were successfully substituted with carboxymethyl groups via esterification. This result is consistent with previous studies [17].

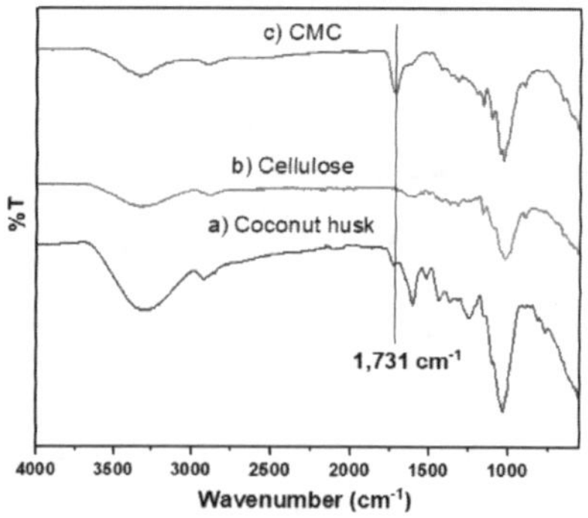


**Figure 3**. FT-IR spectra of a) coconut husk, b) cellulose and c) CMC.

### 3.3 Color of ABA in different buffer solutions

Anthocyanins were extracted from butterfly pea flowers, and their behavior in response to pH changes was investigated. Figure 4 illustrates the color of anthocyanin extracted from butterfly pea flowers (ABA) in different pH buffer solutions (pH 1-14). Buffers with pH values ranging from 1 to 14 were prepared using 1.0 mol/L of HCl solution and 0.5 mol/L of $NH_4OH$ solution (the pH of solutions from 5 to 14 was adjusted by adding small amounts of dilute HCl and $NH_4OH$ solutions). The changes in the color of ABA primarily resulted from alterations in the structure of anthocyanins, as shown in Figure 4. It can be observed that the color of the ABA solutions appeared orange-red when the pH was below 2. This color change occurs because, at low pH, the anthocyanins in the solution predominantly exist in the red flavylium cation structure [18]. From pH 3 to 6, the color shifted from red to light pink due to the flavylium cation structure being disrupted, causing the anthocyanin structure to transition to a neutral quinonoid base [18]. Furthermore, at pH 7-9, the color of ABA solutions transitioned from light pink to blue as the anthocyanin structure shifted to an anionic quinonoid base. At pH values greater than 10, the color of anthocyanins in ABA solutions darkened, progressing from blue to dark green and eventually to yellow. This is because the anthocyanin structure underwent transitions to pale blue quinonoid base structures, green quinonoid base structures, and yellow chalcone structures, respectively [18].

**Figure 4**. Color change of ABA in solutions (pH 1-14).

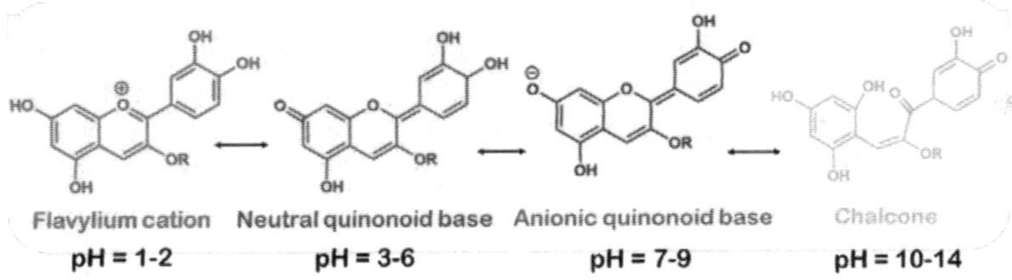


**Figure 5**. Structure change of ABA in solutions (pH 1-14).

### 3.4 Morphology analysis of the smart sensing colorimetric film

The surface morphology of the PVA/CMC films and ABA-PVA/CMC colorimetric film is presented in Figure 6 (images a, b, c, d). As seen in Figure 6 (a), the surface of the PVA/CMC film displays small bubbles, indicating poor compatibility between PVA and CMC. When the ABA content was 0.2 g (0.2-ABA-PVA/CMC), the surface of the film was relatively smooth, suggesting that ABA improved the compatibility between PVA and CMC. However, with an increase in the ABA content, the surfaces of the 0.4-ABA-PVA/CMC and 0.6-ABA-PVA/CMC films developed small bubbles. The high ABA content may have damaged the polymer network structure, making the film structure looser and rougher [19].

IOP Conf. Series: Materials Science and Engineering **1325** (2025) 012005 doi:10.1088/1757-899X/1325/1/012005

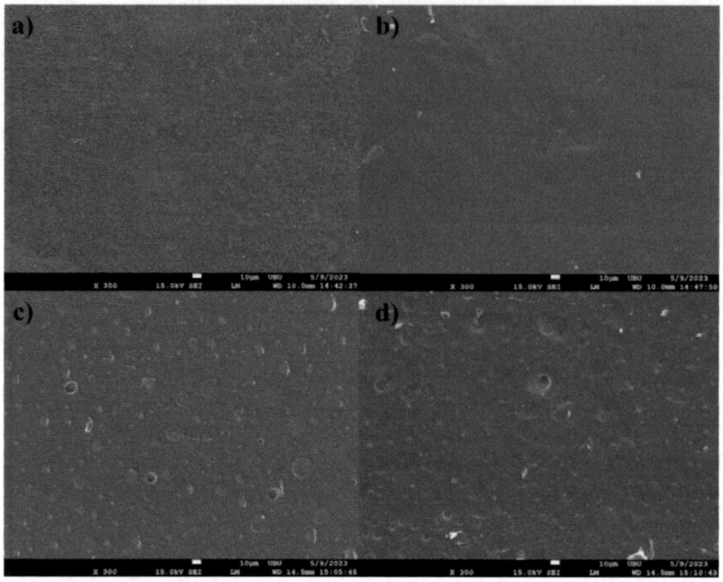


**Figure 6.** Surface SEM image on the cross-sections of PVA/CMA (a), 0.2-ABA-PVA/CMA (b), 0.4-ABA-PVA/CMA (c), and 0.6-ABA-PVA/CMA (d) smart sensing colorimetric film.

### 3.5 Mechanical analysis of the smart sensing colorimetric film

For mechanical testing, the films were molded and cut into dog-bone-shaped tensile specimens according to ASTM D412. Table 1 presents the thickness and tensile strength of all the films. As presented in Table 1, the content of ABA resulted a linear increase in the thickness of the smart sensing colorimetric film, with the 0.6-ABA-PVA/CMC film having the maximum thickness. The mechanical properties test showed that the tensile strength of PVA/CMC films was 20.11 ± 2.21 MPa. With the increase in ABA content, the tensile strength decreased to 19.73 ± 0.29, 16.30 ± 1.93, and 16.09 ± 2.45 MPa for 0.2-ABA-PVA/CMC, 0.4-ABA-PVA/CMC, and 0.6-ABA-PVA/CMC, respectively. The reduction in tensile strength may be attributed to the ABA molecules disrupting the interaction between PVA and CMC molecules [3]. SEM images show that the compatibility of PVA and CMC in the 0.2-ABA-PVA/CMC film was enhanced compared to the PVA/CMC films. However, the 0.2-ABA-PVA/CMC film exhibited lower tensile strength, likely because anthocyanin acts as a plasticizer [18]. Generally, plasticizers can alter the compatibility of blended films [19], which may lead to poor mechanical properties.

**Table 1.** Mechanical properties of colorimetric films

| Colorimetric films | Thickness (mm) | Tensile Strength (MPa) |
|---|---|---|
| PVA/CMA | 0.065 ± 0.02 | 20.11 ± 2.21 |
| 0.2-ABA-PVA/CMA | 0.053 ± 0.03 | 19.73 ± 1.29 |
| 0.4-ABA-PVA/CMA | 0.070 ± 0.01 | 16.30 ± 1.93 |
| 0.6-ABA-PVA/CMA | 0.090 ± 0.02 | 16.09 ± 2.45 |

IOP Conf. Series: Materials Science and Engineering **1325** (2025) 012005 doi:10.1088/1757-899X/1325/1/012005

### 3.6 The ABA-PVA/CMC film's response to pH changes

The pH sensitivity is an key evaluation criterion for intelligent packaging used to monitor the freshness of high-protein foods such as meat, dairy, and seafood. As shown in Table 2, different colors were observed with changes in pH, which were used to assess the sensitivity of the colorimetric films to ammonia and acetic acid solutions. These solutions represent the decomposition of high-protein foods such as meat and seafood. It can be seen that the color of the 0.2-ABA-PVA/CMC film did not show a clear color change after immersion in NH4OH and CH3COOH solutions. This is because the low content of anthocyanins in the polymer matrix, which limits the intermolecular interactions between anthocyanins and biopolymers, thereby restricting the structural changes in anthocyanins [8]. As the content of ABA powder increased, the 0.4-ABA-PVA/CMC and 0.6-ABA-PVA/CMC films showed a clear color change. Therefore, the smart sensing colorimetric films were able to monitor pH variations, exhibiting different color due to structural alterations in response to various acidic and basic environments. In the pH change test, $NH_4OH$ interacted with the smart sensing colorimetric films, creating an alkaline environment that induced a shift in the anthocyanin structure from the red flavylium cation (yellow salt ion) to the blue quinone base and light yellow chalcone. Consequently, the color of the colorimetric films gradually changed to light green. As presented in Table 2, the 0.4-ABA-PVA/CMC film exhibited the highest sensitivity to the ammonia solution. In the acidic pH change test, acetic acid reacted with the smart sensing colorimetric films to produce an acidic environment, leading to a structural change in the anthocyanins from the red flavylium cation (yellow salt ion) to the pseudobase form [18]. As a result, the color of the smart sensing colorimetric films gradually faded to colorless, as presented in Table 2.

**Table 2.** PVA/CMC-ABA film' response to pH changes in $NH_4OH$ and $CH_3COOH$ solution

| Colorimetric films | Color | $NH_4OH$ | $CH_3COOH$ |
|---|---|---|---|
| | | Color response | Color response |
| PVA/CMA |  |  |  |
| 0.2-ABA-PVA/CMA |  |  |  |
| 0.4-ABA-PVA/CMA |  |  |  |
| 0.6-ABA-PVA/CMA |  |  |  |

Therefore, the optimal response of the PVA/CMC-ABA film to pH changes in $NH_4OH$ and $CH_3COOH$ solutions was observed in the 0.4-ABA-PVA/CMC film, as it exhibited a clear color change along with acceptable mechanical properties. It is widely recognized that gas sensors should exhibit a rapid response to analytes. Therefore, the 0.4-ABA-PVA/CMC film, which showed the most significant color change, shows great potential for real-time freshness monitoring of food..

## 4. Conclusions

In this work, novel smart sensing colorimetric films were successfully developed by incorporating varying amounts of anthocyanin extracted from butterfly pea flowers (ABA) into a poly(vinyl alcohol)/carboxymethyl cellulose (PVA/CMC) matrix. Coconut husks were treated with NaOH and $H_2O_2$ for bleaching before undergoing carboxymethylation. The bleached coconut husks were converted into CMC through alkalization, followed by etherification using NaOH and $ClCH_2COOH$ in an alcoholic medium. SEM analysis revealed that the addition of ABA enhanced the compatibility between PVA and CMC. However, the inclusion of ABA resulted in a reduction in tensile strength. Notably, films with a PVA/CMC ratio of 4:1 exhibited the highest tensile strength and uniform dispersion. Furthermore, ABA-PVA/CMC films demonstrated strong pH sensitivity, undergoing distinct color changes at various pH levels. Application tests confirmed that the colorimetric films could effectively indicate food freshness, specifically by testing their sensitivity to ammonia and acetic acid solutions. These colorimetric films hold significant potential for reducing the risk of consuming spoiled food by monitoring freshness in real-time.

## Acknowledgement

The research is gratefully acknowledged and financially provided by the Department of Chemistry and the Faculty of Science, Ubon Ratchathani University.

## References

[1] Kuswandi B and Nurfawaidi A 2017 *Food Control.* **82** 91-100.
[2] Freitas P A, Silva R R, de Oliveira T V, Soares R R, Junior N S, Moraes A R, Pires S C and Soares N F 2020 *LWT.* **132** 109780.
[3] Liu D, Cui Z, Shang M and Zhong Y 2021 *Food Packag. Shelf Life.* **28** 100641.
[4] Liu D, Zhong Y, Pu Y, Li X, Chen S and Zhang C 2023 *Foods.* **12** 2316.
[5] Kuswandi B, Restyana A, Abdullah A, Heng L Y and Ahmad M 2012 *Food Control.* **25** 184-189.
[6] Zhang J, Zou X, Zhai X, Huang X, Jiang C and Holmes M 2019 *Food Chem.* **272** 306-312.
[7] Merz B, Capello C, Leandro G C, Moritz D E, Monteiro A R and Valencia G A 2020 *Int. J. Biol. Macromol.* **153** 625-632.
[8] Yong H and Liu J 2020 *Food Packag. Shelf Life.* **26** 100550.
[9] Havananda T and Luengwilai K 2019 *Genet Resour Crop Evol.* **66** 645-658.
[10] Poh N. A. B. A. 2019 *School of Science and Technology: Singapore.*
[11] Wang Y, Zhang J and Zhang L 2022 *Food Chem.* **373** 131367.
[12] Peng W, Wang L, Ohkoshi M A and Zhang M 2015 *Cellul. Chem. Technol.* **49** 756-764.
[13] Li B, Xu W, Kronlund D, Määttänen A, Liu J, Smått J H and Xu C 2015 *Carbohydr. Polym.* **133** 605-612.
[14] Caputo D, Fusco C, Nacci A, Palazzo G, Murgia S, D'Accolti L and Gentile L 2021 *Carbohydr. Polym. Technol. Appl.* 2, 100094.
[15] Astruc J, Nagalakshmaiah M, Laroche G, Grandbois M, Elkoun S and Robert M 2017 *Carbohydr. Polym.* **178** 352-359.
[16] Zhang X, Huang H, Qing Y, Wang H and Li X 2020 *Materials.* **13**.
[17] Atta O M, Manan S, Ul-Islam M, Ahmed A A Q, Ullah M W and Yang G 2022 *Adv. Compos. Hybrid Mater.* 1-18.
[18] Castañeda-Ovando A, de Lourdes Pacheco-Hernández M, Páez-Hernández M E, Rodríguez J A and Galán-Vidal C A 2009 *Food Chem.* **113** 859-871.
[19] Zhai X, Shi J, Zou X, Wang S, Jiang C, Zhang J and Holmes M 2017 *Food Hydrocoll.* **69** 308-317.

IOP Conf. Series: Materials Science and Engineering **1325** (2025) 012006 doi:10.1088/1757-899X/1325/1/012006

# Development of polylactic acid composite enhance with cassava pulp and cassava strach

**B Ketnapha[1], F Saranyakan[1], W Kittiya[1] and J Rukkiat[1,*]**

[1] Department of Chemistry, Faculty of Science, Ubon Ratchathani University, Thailand.

*Corresponding author: rukkiat.j@ubu.ac.th

**Abstract.** In this study, we investigated the preparation of polylactic acid (PLA) composites enhanced with environmentally friendly cassava pulp (CP) and cassava starch (CS). The study aimed to optimize the blend ratio of these cassava-based materials with PLA to improve mechanical properties and biodegradability. Six biocomposites were prepared with a fixed 40% PLA content and CP(g) to CS(g) ratios of 1.0:1.5, 1.0:1.2, 1.0:1.0, 1.2:1.0, and 1.5:1.0. The addition of CP enhanced the composites' toughness and hardness, with tensile strength increasing from 3.56 MPa at 0% CP to 6.96 MPa at 26.53% CP. However, TGA results indicated a reduction in thermal stability with higher CP content. SEM images showed a homogeneous distribution of TPS and CP, though surface roughness became more pronounced at increased CP levels, leading to visible phase separation between CP and CS, which likely contributed to the observed reduction in thermal stability. In conclusion, PLA/CS/CP biocomposites exhibit effective molding capabilities and represent a more sustainable material option for various applications.

## 1. Introduction

Plastic waste is one of the most critical environmental challenges due to its persistence and non-biodegradability. According to the Pollution Control Department of Thailand, municipal plastic waste amounted to approximately 60,000 tons per day in 2012 [1], a sharp increase from the previous year. Improper disposal and accumulation of plastic waste can cause severe environmental problems, including contamination of soil and water, the release of toxic substances during incineration, and contributions to global warming through greenhouse gas emissions.

Bioplastics, such as polylactic acid (PLA), have gained attention as sustainable alternatives to conventional plastics. Derived from renewable resources like corn and cassava starch, PLA is biodegradable and environmentally friendly. However, PLA's inherent brittleness and low thermal stability limit its application in flexible or durable materials [2].

Thermoplastic starch (TPS), derived from plant starch with the addition of plasticizers like glycerol, is another promising bioplastic. TPS is biodegradable, cost-effective, and derived from abundant resources, such as cassava [3]. However, it is limited by its mechanical weakness, sensitivity to moisture, and poor thermal stability.

IOP Conf. Series: Materials Science and Engineering **1325** (2025) 012006 doi:10.1088/1757-899X/1325/1/012006

Agricultural waste is generated in millions of tons annually, and its improper management and disposal result in severe environmental impacts, including ecosystem degradation [4]. Cassava pulp (CP), a major byproduct of cassava starch production, serves as a notable example of agricultural waste. In Thailand, particularly in Ubon Ratchathani province, cassava pulp is abundantly available; however, its utilization remains limited, posing significant challenges related to environmental pollution and waste disposal. Effective management of this waste has the potential not only to mitigate environmental harm but also to create opportunities for the development of sustainable materials.

To overcome these limitations, blending PLA with TPS and CP, a byproduct of cassava starch production, has been explored. Cassava pulp contains residual starch and fibers, making it a potential reinforcement material for biocomposites. By combining these components, it is possible to develop biocomposites with improved mechanical and thermal properties while reducing agricultural waste.

## 2. Experimental

### 2.1 Materials

Cassava starch was purchased from CP Axtra Public Company Limited, Thailand, while cassava pulp (CP) was provided by Ubon Bio Ethanol in Ubon Ratchathani Province, Thailand. Poly (lactic acid) was obtained from NatureWorks Asia Pacific Company Limited, Thailand. Citric acid was purchased from Ajax Finechem Pty Limited, Germany. Chloroform, glycerol, hydrochloric acid, and ethanol were procured from Dasit Group, Italy.

### 2.2 Preparation of cassava pulp

Cassava pulp powder was prepared by first thoroughly washing the cassava pulp to remove impurities. The cleaned pulp was then dried at 60°C for 12 hours to eliminate excess moisture. The dried material was finely ground using a herb blender operating at a speed of 25,000 revolutions per minute. The resulting powder was subsequently sieved through a 40-mesh sieve to ensure uniform particle size and remove any remaining large particles or impurities.

### 2.3 Preparation of TPS/PLA and TPS/PLA/CP biocomposite

The biocomposites were prepared using an internal mixer, according to the formulations outlined in Table 1. Initially, cassava starch, cassava pulp, glycerol (as a plasticizer), and citric acid (as a crosslinking agent) were blended to create a thermoplastic starch (TPS) mixture. This TPS mixture was then combined with PLA in the internal mixer at 175°C for 15–20 minutes to achieve a homogeneous composite. After mixing, the composite was molded into sheets using compression molding or a two-roll mill, then cooled and stored in sealed containers.

**Table 1.** Biocompositions of six formulations with varying weight ratios of cassava pulp (CP) and cassava starch.

| Formula | PLA (wt%) | Cassava Starch (wt%) | Cassava Pulp (wt%) | Starch:CP Ratio | Glycerol (wt%) | Citric Acid (wt%) |
|---|---|---|---|---|---|---|
| TPS/PLA | 40 | 44.22 | 0 | - | 15.50 | 0.30 |
| TPS/PLA/CP1 | 40 | 26.53 | 17.69 | 1.5:1.0 | 15.50 | 0.30 |
| TPS/PLA/CP2 | 40 | 25.80 | 18.43 | 1.2:1.0 | 15.50 | 0.30 |
| TPS/PLA/CP3 | 40 | 22.11 | 22.11 | 1.0:1.0 | 15.50 | 0.30 |
| TPS/PLA/CP4 | 40 | 18.43 | 25.80 | 1.0:1.2 | 15.50 | 0.30 |
| TPS/PLA/CP5 | 40 | 17.69 | 26.53 | 1.0:1.5 | 15.50 | 0.30 |

### 2.4 Characterization and property testing of TPS/PLA and TPS/PLA/CP biocomposites

2.4.1 Tensile strength Test

The tensile test specimens were prepared in accordance with the ISO 527-2:2012 standard [5], specifically using Type 1B specimens, and tested using a universal testing machine.

2.4.2 FT-IR characterization

Fourier transform infrared spectroscopy (FTIR) was used to analyze the chemical groups of cassava pulp. The analysis was performed in attenuated total reflection mode (ATR-FTIR), with spectra recorded over a wavenumber range of 4000–450 cm$^{-1}$.

2.4.3 Thermal stability

The thermal stability of the biocomposites was evaluated using thermogravimetric analysis (TGA). The specimens were analyzed under a nitrogen atmosphere at a flow rate of 20 mL/min, over a temperature range of 30–600°C, with a heating rate of 10°C/min.

2.4.4 Scanning electron microscopy (SEM)

Scanning electron microscopy (SEM) was employed to capture detailed images of the biocomposite surfaces, providing valuable insights into their surface morphology. Furthermore, the SEM analysis facilitated the assessment of the size and distribution of TPS and CP within the samples.

## 3. Results and discussion

### 3.1 Preparation of cassava pulp

Cassava pulp, sourced from a cassava starch factory (Ubon Bio Ethanol) in Ubon Ratchathani Province, is shown in Figure 1a. The dried materials were then ground into a powder, resulting in a white-brown cassava pulp powder, as shown in Figure 1b.

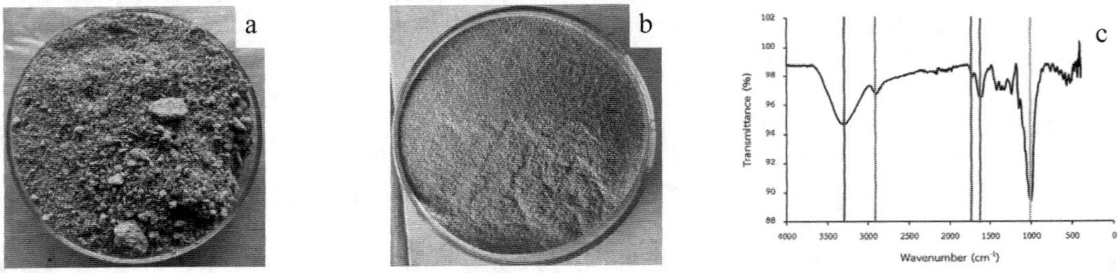


**Figure 1.** Photographs of (a) cassava bagasse, (b) cassava bagasse powder (CP), and (c) the FTIR spectrum of cassava bagasse powder.

The FTIR spectroscopy analysis of cassava pulp, shown in Figure 1c, revealed a peak for C=O stretching at 1627 $cm^{-1}$ (red), which indicates the presence of hemicellulose. A C=C stretching peak at 1605 $cm^{-1}$ (green) corresponds to lignin, while a C-O stretching peak appears at 1033 $cm^{-1}$ [4]. Additionally, significant peaks were observed, including O–H stretching at 3325.7 $cm^{-1}$ (purple) and C–H stretching of alkene at 2922 $cm^{-1}$ (blue), which indicate the presence of cellulose [6]. The FTIR spectrum shows that cassava bagasse contains a combination of lignin, hemicellulose, and cellulose.

### 3.2 Morphological characteristics of TPS/PLA and TPS/PLA/CP biocomposites by SEM

SEM micrographs of the fracture surfaces of TPS/PLA and TPS/PLA/CP biocomposites, shown in Figure 2, confirm these observations. The PLA surface (Figure 2a) exhibits a smooth texture without any signs of matrix deformation. In contrast, the surface morphology of TPS/PLA, as shown in Figure 2b, reveals a rougher texture with small spherical droplets scattered across the surface, indicating a lack of continuity between the TPS and PLA phases. When cassava bagasse was added to TPS/PLA/CP1 (Figure 2c), the surface became smoother, and the number of small spherical droplets decreased, signifying improved phase compatibility facilitated by the addition of cassava bagasse. With further additions of cassava bagasse, as shown in Figures 2 (d–f), the surface became progressively smoother and more uniform. At the highest cassava bagasse content, as depicted in Figure 2g, no spherical droplets were observed, indicating excellent phase compatibility and complete integration of cassava bagasse into the matrix.

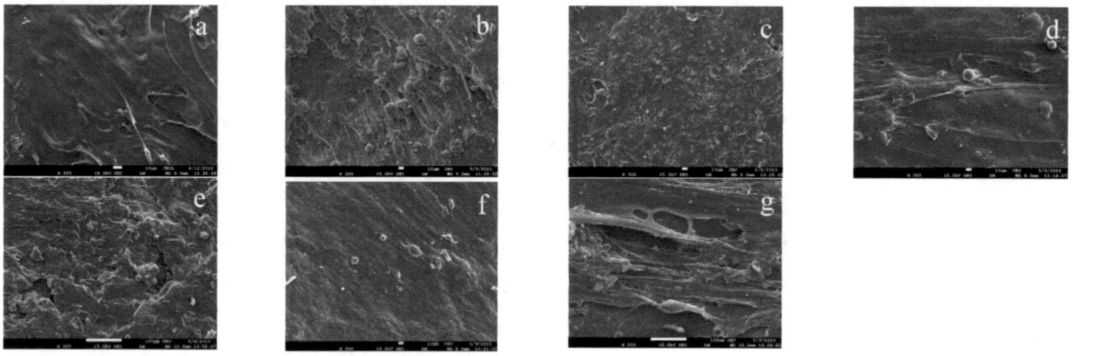


**Figure 2.** SEM micrographs of the fracture surfaces of pure PLA (a), TPS/PLA (b), TPS/PLA/CP1 (c), TPS/PLA/CP2 (d), TPS/PLA/CP3 (e), TPS/PLA/CP4 (f), and TPS/PLA/CP5 (g) biocomposite.

To examine the phase structure of the materials, the compatibility between the PLA and TPS phases, and the distribution and size of the dispersed phase, a selective extraction was performed on all cryofractured samples. Four of the six samples were selected for this experiment: TPS/PLA, TPS/PLA/CP1, TPS/PLA/CP4, and TPS/PLA/CP5. Chloroform ($CHCl_3$) solution was used as solvent to selectively remove the PLA phases. The surface morphologies of the etched samples after extraction were then analyzed and are presented in Figures 3B.

The TPS/PLA blend exhibited a smooth cryofractured surface with small spherical droplets dispersed within the matrix (Figure 3A (a)). However, after extraction with $CHCl_3$, no spherical droplets were observed in the specimen (Figure 3B (a)), confirming that the droplets were originally part of the PLA phase. The removal of PLA using chloroform significantly affects the morphological characteristics of the biocomposites. Upon the addition of cassava pulp, as shown in Figure 3B (b), the phase exhibits an elongated structure, which indicates the initial transformation of the phase morphology. With increasing concentrations of cassava pulp, as observed in Figures 3B (c) and 3B (d), the phase structure transitions to a more continuous form, accompanied by the development of wider gaps. This progression highlights the improved compatibility of the blend, facilitated by the incorporation of cassava bagasse. The results clearly demonstrate that the addition of cassava pulp enhances the interaction between the composite phases, effectively converting the phase morphology from discontinuous to continuous.

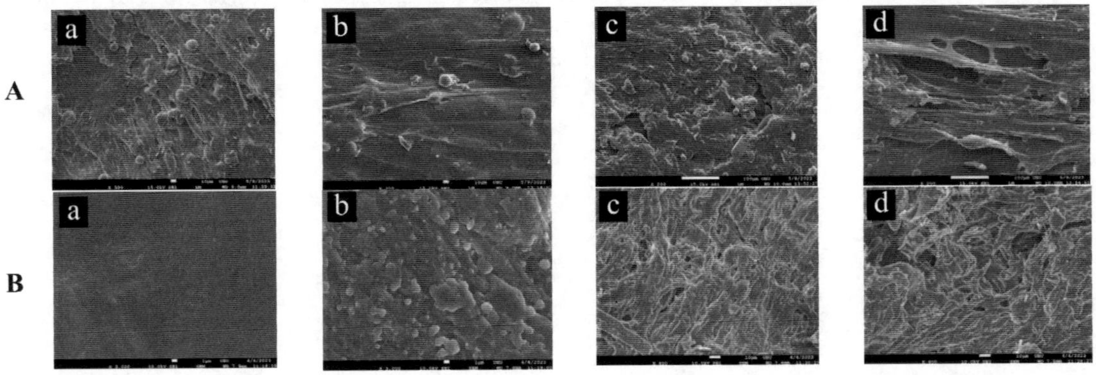


**Figure 3.** SEM micrographs of (A) cryofractured surface, and (B) cryofractured surface extracted with chloroform for (a) TPS/PLA, (b) TPS/PLA/CP1, (c) TPS/PLA/CP4, and (d) TPS/PLA/CP5 biocomposite.

### 3.3 Morphological characteristics of TPS/PLA and TPS/PLA/CP biocomposites by Stereo microscope (SM)

The results of the morphological study of the pure PLA formulation, which contains no additives, are shown in Figure 4a, presenting an SM image of pure PLA. The image demonstrates that no additional particles are present in the polylactic acid matrix, except for dispersed air bubbles. The material exhibits a clear and slightly translucent appearance. In contrast, the morphological study of the TPS/PLA blend, shown in Figure 4b, reveals that cellulose particles are well dispersed within the polylactic acid matrix, indicating good compatibility between the components. This blend appears opaquer while maintaining slight translucency compared to pure PLA. Finally, the morphological study of the TPS/PLA/CP composite, displayed in Figure 4c, shows that cellulose particles are also well dispersed throughout the polylactic acid matrix. Additionally, glycerol droplets are observed scattered across the surface of the composite, likely due to residual moisture from the addition of glycerol as a plasticizer. The TPS/PLA/CP composite exhibits a significantly opaquer appearance than the TPS/PLA blend, suggesting enhanced dispersion and better incorporation of the components.

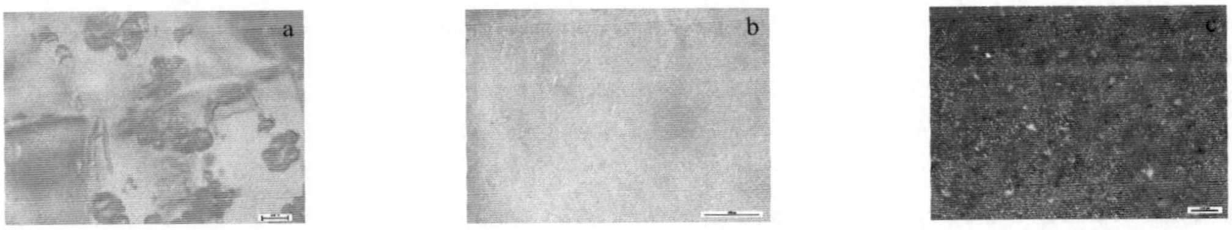


**Figure 4.** SM micrographs of (a) pure PLA, (b) TPS/PLA, and (c) TPS/PLA/CP1 biocomposite.

### 3.4 Tensile and Thermal properties of TPS/PLA and TPS/PLA/CP biocomposites

Figure 5a illustrates the tensile strength of TPS/PLA, TPS/PLA/CP1, TPS/PLA/CP2, TPS/PLA/CP3, TPS/PLA/CP4, and TPS/PLA/CP5 biocomposites. The results reveal that the TPS/PLA blend exhibited poor tensile strength due to the addition of TPS, which negatively affected the composite's elasticity, as previously reported by [7]. However, the incorporation of cassava pulp (CP) into TPS/PLA significantly enhanced the tensile strength and stiffness, with these properties further improving at higher CP concentrations [8]. This enhancement was attributed to the reinforcing effect of CP fibers and improved interfacial compatibility between the TPS and PLA phases, as evidenced by SEM analysis. The interaction between cassava starch (CS) in TPS, functioning as the matrix, and the fibrous CP, acting as a reinforcing filler, was crucial in improving the composite properties. The cellulosic content of CP facilitated better interfacial adhesion, reduced phase separation, and enhanced load transfer within the composite. Consequently, the synergy between CS and CP resulted in improved mechanical performance, particularly in biocomposites with higher CP content.

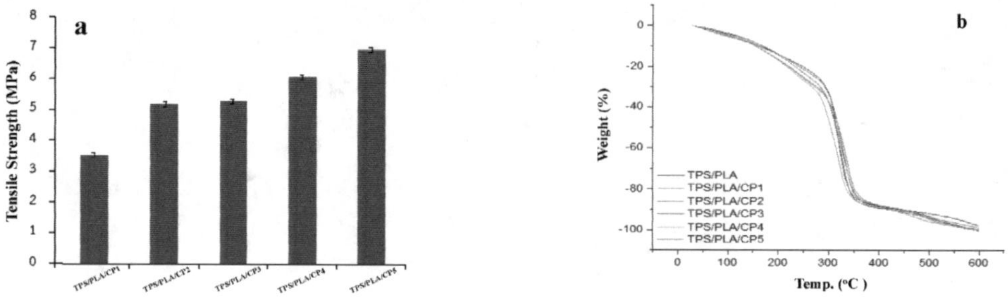


**Figure 5.** Tensile strength (a) and TGA curves (b) of the TPS/PLA, TPS/PLA/CP1, TPS/PLA/CP2, TPS/PLA/CP3, TPS/PLA/CP4, and TPS/PLA/CP5 biocomposite.

The thermal stability of TPS/PLA and TPS/PLA/CP composites was investigated using thermogravimetric analysis (TGA). Figure 5b illustrates the weight loss of each material component as a function of temperature. The TPS/PLA blend exhibited two distinct stages of weight loss: the first stage, occurring between 130–170°C, corresponded to the evaporation of water and glycerol, while the second stage, between 220–385°C, represented the thermal decomposition of TPS and PLA, respectively. In contrast, the TPS/PLA/CP biocomposites displayed three stages of weight loss. The first stage, at 130–180°C, was attributed to the evaporation of water and glycerol. The second stage, at 182–228°C, corresponded to the initial decomposition of TPS, while the third stage, at 260–390°C, was associated with the decomposition of PLA. As shown in Figure 5b, TPS/PLA/CP samples containing cassava pulp exhibited more pronounced weight loss, with a steady decline in the graph. The TGA results indicated that thermal stability decreased with increasing cassava pulp content. Furthermore, the lowest decomposition temperature, 321.94°C, was recorded for the sample with the highest cassava pulp concentration, highlighting the impact of CP on the thermal stability of the biocomposites. This reduction in thermal resistance may be attributed to the higher plasticizing effect of cassava pulp, which weakens the strong intermolecular bonds among polymer chains. A similar trend was observed in glycerol-plasticized sugar palm starch, where increasing glycerol concentration reduced thermal resistance [9].

## 4. Conclusions

In this study, we successfully developed biocomposites by blending PLA with TPS and cassava pulp (CP) to enhance their mechanical and thermal properties while addressing agricultural waste. The results demonstrated that incorporating cassava pulp significantly improved the tensile strength and stiffness of the TPS/PLA blends. These improvements were attributed to the reinforcing effect of CP fibers and the enhanced compatibility between the TPS and PLA phases, as confirmed by SEM analysis. Thermal stability analysis revealed that the addition of CP reduced the composites' thermal stability, likely due to its plasticizing effect. FTIR analysis further confirmed the successful integration of cassava pulp into the biocomposite matrix. Overall, these findings highlight the potential of cassava pulp as a sustainable filler in biocomposites, offering an effective solution for reducing agricultural waste and providing an environmentally friendly alternative to conventional plastics. Further optimization of these biocomposites could broaden their applications across various industries, including packaging, automotive, and construction, where the demand for biodegradable and eco-friendly materials is rapidly increasing. The successful incorporation of agricultural waste into high-performance biocomposites not only supports waste reduction but also contributes to the development of sustainable materials, fostering a circular economy and minimizing the environmental impact of plastic use.

**Acknowledgement**
I would like to thank the Faculty of Science, Ubon Ratchathani University, for their support in providing equipment and materials, which contributed to the successful completion of this research.

**References**
[1] https://www. researchgate.net/profile/AthanasiosGlobal-Municipal-Solid-Waste-MSW-in-Crisis-Two-billion-tonnes-of-MSW-every-year-a-worrying-worldwide-environmental-problem.pdf; November 2024.
[2] Tripathi N, Misra M and Mohanty A-K 2021 *J. ACS Eng. Au.* **1** 7.
[3] Diyana Z-N, Jumaidin R, Selamat M-Z, Ghazali I, Julmohammad N, Huda N and Ilyas R-A 2021 *Polymers*. **13** 1396.
[4] Mujtaba M, Fraceto L-F, Fazeli M, Mukherjee S, Savassa S-M, de Medeiros G-A and Vilaplana F 2023 *J. Clean. Prod.* **402** 136815.
[5] https://www.cabidigitallibrary.org/doi/full/10.5555/20063078259; November 2024.
[6] Lin Q, Huang Y and Yu W 2021 *Ind. Crops Prod.* **169** 113640.
[7] Heidemann H-M, Dotto M-E, Laurindo J-B, Carciofi B-A and Costa C 2019 *Colloids Surf. A Physicochem. Eng. Asp.* **580** 123739.
[8] Jullanun P and Yoksan R 2020 *Polym. Testing* **88** 106522.
[9] Sanyang M-L, Sapuan S-M, Jawaid M, Ishak M-R and Sahari J 2015 *Polymers*. **7**.

IOP Conf. Series: Materials Science and Engineering **1325** (2025) 012007 doi:10.1088/1757-899X/1325/1/012007

# Development of the thermoplastic vulcanizate material between bromobutyl rubber and polypropylene: compatibility study

**Siriwan Klinjan[1, 2], Supansa Khunkayan[2], Peerut Suwanpimolkul[2], Supitta Suethao[1,3], and Wirasak Smitthipong [1,3,*]**

[1]Specialized center of Rubber and Polymer Materials in agriculture and industry (RPM), Department of Materials Science, Faculty of Science, Kasetsart University, Bangkok, 10900, Thailand
[2]S.K. Polymer Co.,Ltd., Bangkok, 10150, Thailand
[3]Hub of Talents in Natural Rubber, National Research Council of Thailand (NRCT), Bangkok, 10900, Thailand

*Corresponding author: fsciwssm@ku.ac.th

**Abstract**. Thermoplastic vulcanizate (TPV), which is based on bromobutyl rubber (BIIR) and polypropylene (PP), combines the good melt-processability and recyclability of PP with its excellent elasticity. In the structure of BIIR, there are a few bromines that have a polar component of about 1.8%. When BIIR is mixed with PP, it causes poor compatibility due to a non-polar PP. This research aims to prepare TPV from BIIR and PP by the melt mixing process. The ratio of rubber to plastic was set at 65:35. After that, to increase the compatibility, polypropylene-grafted maleic acid anhydride (PP-g-MA) was used as an enhancer at different concentrations. The mechanical properties of the TPV with and without a filled compatibilizer were studied. The results demonstrated that adding PP-g-MA considerably enhanced the tensile strength, suggesting improved BIIR and PP compatibility. By incorporating PP-g-MA also resulted in an increase in tear resistance, which suggests enhanced adhesion between BIIR and PP. Furthermore, all TPVs maintain a similar hardness Shore A. This work could be applied to some specific TPV applications, especially for the recycle TPV products.

## 1. Introduction

High temperature polymers recently become the fastest growing elastomer to replace non-recyclable petroleum-based rubbers. Brominated butyl rubber (BIIR) is an important brominated copolymer product of a butyl rubber, which shows excellent gas resistance. It also demonstrates high resistance to heat, acids and alkalis that results butyl rubber with a faster melting speed that causes good internal adhesion as well as with other materials. This is due to the fact that BIIR chains contain 1.9–2.1% of polar bromine derivatives with adhesive effect when mixed with other immiscible plastic materials. Therefore, it is essential to dynamically mix the compatibilizers in the vulcanization process to increase the compatibility between the polymer chains of rubber and plastic [2]. The results showed that the tensile strength of the compatible mixture butyl rubber with polypropylene was significantly higher than that of the incompatible mixture, while percentage elongation and Young's modulus were comparable. The morphology of the mixtures was examined by scanning electron microscopy (SEM) and it revealed that the compatible mixtures exhibited a marked reduction in the dispersed domain size. Incompatible blends of epoxidized natural rubber (ENR) and polypropylene (PP) can be ameliorated using PP-g-MA providing the ENR/PP TPV with superior mechanical properties [4]. In the present research, PP-g-MA was applied as a compatibilizer between BIIR rubber and PP to study its effect on various properties such as mechanical properties of TPV for the functional applications.

## 2. Experimental

### *2.1 Materials*

Bromobutyl rubber (BIIR) was supplied by ARLANXEO, Maastricht, Netherlands, Polypropylene random copolymer (PP) was supplied by HMC Polymers, Bangkok, Thailand, Anhydride modified polypropylene (PP-g-MA) was supplied by Dow, Rayong, Thailand, Dicumyl peroxide (DCP) was supplied by Thai Rubb Tech, Bangbon Bangkok, Thailand and trimethylolpropane trimethacrylate (TMPTMA) was supplied by Thai Rubb Tech, Bangbon Bangkok, Thailand.

### *2.2 Preparation of TPV*

PP and A were melted for 5 min using the internal mixer at 165°C and 10 rpm. For formulation with compatibilizer, PP-g-MA was added in the first step to melt along with PP and A. BIIR was then added and mixed for 5 min using the internal mixer at the same condition. After that, compounds were cooled down in a thin sheet with a two-roll mill, this type of sample was called Thermoplastic polyolefins (TPO). The TPO was masticated to become softer before adding the curing agents for 2 min using the internal mixer at 60°C and 30 rpm and cooled down again in a thin sheet.

**Table 1.** The formulations of TPV in phr (parts per hundred of rubber).

| Formula | G0 | G1 | G2 | G3 | G4 |
|---|---|---|---|---|---|
| 1. BIIR | 65 | 65 | 65 | 65 | 65 |
| 2. A | 25 | 25 | 25 | 25 | 25 |
| 3. PP | 10 | 9 | 8 | 7 | 6 |
| 4. PP-g-MA | - | 1 | 2 | 3 | 4 |
| 5. Vulcanizing agent | 1.5 | 1.5 | 1.5 | 1.5 | 1.5 |
| 6. Accelerator | 1 | 1 | 1 | 1 | 1 |

G means anhydride modified polypropylene (PP-g-MA)
A means melting processing aids

### *2.3 Characterizations*

The tensile strength was determined by universal testing machine (LLOYD, Thailand) with the tensile rate of 500 mm/min measured following JIS K 6301 Type 2. The tear strength measured following ASTM D 624 (Die C) by taking a sample sheet approximately 2 mm thick and cutting it with an angled test piece cutter, pull rate 500 mm/min, calculated according to equation (1)

$$T_S = \frac{F}{t} \quad (1)$$

where $T_S$ is the tear strength (kgf/cm²), F is the force that causes the break, and t is thickness.
The hardness of TPV was tested by Shore A following JIS K 6253 Type A using Durometer (Tecklock GS-610, Japan), three specimens of each sample were tested and the then average value was taken. Compatibility was also evaluated by dynamic mechanical analyzer (Mettler-Toledo, Switzerland).

## 3. Results and discussion

The tensile strength of all samples (G0, G1, G2, G3, and G4) show the values between 35-50 kgf/cm2 (Figure 1) which increases with the addition of compatibilizer, indicating improved compatibility between BIIR and PP. This improvement is clearly caused by the addition of PP-g-MA cross-linker.

For the stress at 100% and 300% Modulus (Table 2.), there was also an increasing trend. This further confirms the ability of the compatibilizer to improve the compatibility between BIIR and PP, resulting in increased material strength.

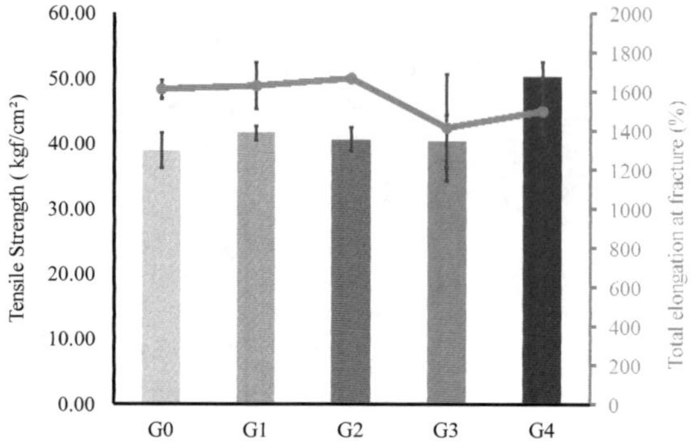


**Figure 1.** The tensile strength and percentage total elongation at fracture of different TPV samples.

**Table 2.** The modulus of TPV (kgf/cm$^2$)

| Sample | Modulus at 100% (kgf/cm$^2$) | Modulus at 300% (kgf/cm$^2$) |
|---|---|---|
| G0 | 6.32 ± 0.63 | 10.62 ± 1.00 |
| G1 | 9.04 ± 1.44 | 14.51 ± 2.17 |
| G2 | 7.95 ± 0.52 | 12.98 ± 0.68 |
| G3 | 10.25 ± 2.04 | 16.90 ± 2.71 |
| G4 | 12.23 ± 0.87 | 20.38 ± 1.46 |

For the tear strength of all samples (G0, G1, G2, G3 and G4), it can be seen that the results tended to increase when PP-g-MA was added (Figure 2). The reason was that MA interacted with Br in the BIIR structure, so the tear strength tended to increase significantly.

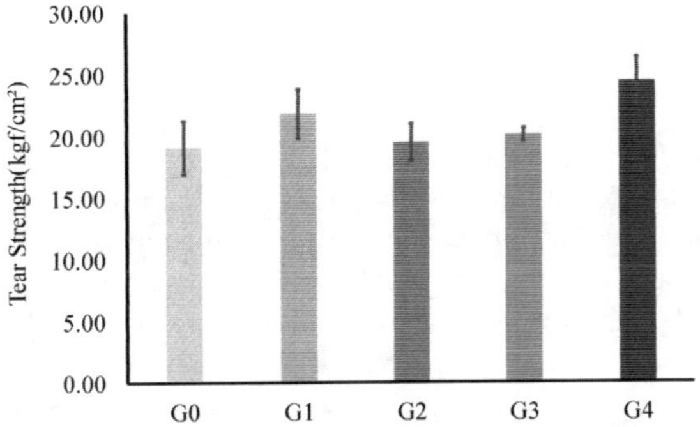


**Figure 2.** The tear strength of the TPV samples.

The hardness shore A of G0, G1, G2, and G3 samples were almost similar (Figure 3), or there were no significantly different of hardness of samples with PP-g-MA below 4 phr, indicating that PP-g-MA did not affect the hardness of TPV because the structure of PP-g-MA was close to PP and had a low MA ratio.

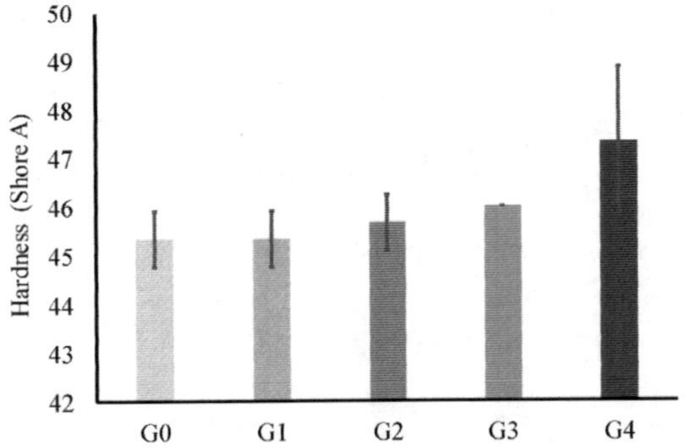


**Figure 3.** The hardness of TPV samples.

The compatibility between BIIR and PP was also tested by a dynamic mechanical analyzer (DMA). From the rubbery plateau range (Figure 4.), it was found that the storage modulus (E') tended to increase with the addition of PP-g-MA due to the increased compatibility between BIIR and PP. Tan delta or dissipation energy was reduced with increasing of the compatibilizer to connect BIIR and PP becoming like a network. This was in good agreement with the previous work [5], higher network of rubber exhibited lower dissipation energy.

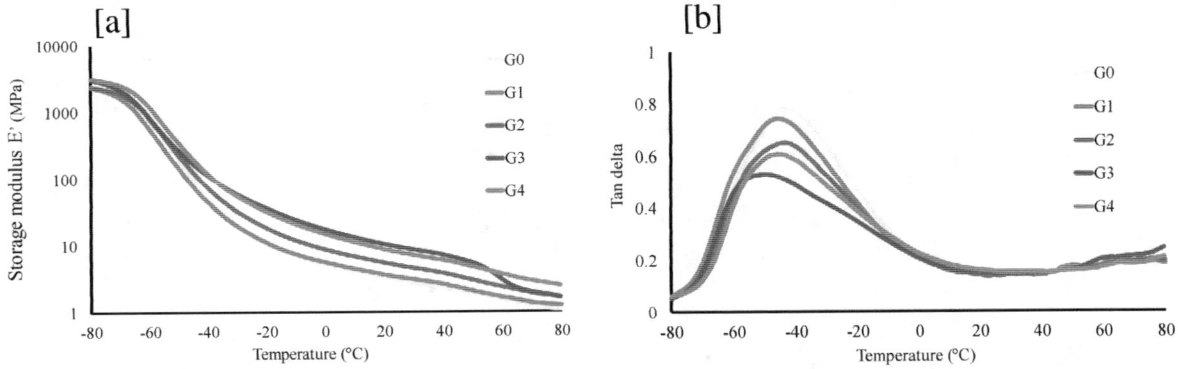


**Figure 4.** Storage modulus E' (MPa) [a], and tan δ [b] of TPV samples.

IOP Conf. Series: Materials Science and Engineering **1325** (2025) 012007 doi:10.1088/1757-899X/1325/1/012007

## 4. Conclusions

In this research, PP-g-MA was used as a compatibilizer to improve the structure of TPVs fabricated between bromobutyl rubber (BIIR) and polypropylene (PP). A comparison of the mechanical properties of TPVs with and without PP-g-MA showed an increase in tensile and tear strengths, which was attributed to the increased compatibility. The results of DMA also confirmed this compatibility, while the hardness of TPVs was not significantly different below 4 phr of PP-g-MA. Finally, beyond 2 phr of PP-g-MA were the compromised concentration for BIIR/PP blend depending on the functional applications of TPV.

## Acknowledgement

S.K. Polymer Co.,Ltd., Bangkok, Thailand and Specialized Center of Rubber and Polymer Materials in agriculture and industry (RPM), Department of Materials Science, Faculty of Science, Kasetsart University, Bangkok, Thailand for finantial supports.

## References

[1] J. Hou, M. Zhong, X. Pan, L. Chen, X. Wu, Z. Kong, *et al.* Composites Part B: Engineering, 2020, 108220.
[2] Q. Tang, S. Hu, L. Han, C. Zong and J. Sun. Polymers, 2022, 2740.
[3] S. Chongprakobkit, M. Opaprakasit and S. Chuayjuljit. Journal of Metals, Materials and Minerals, 2007.
[4] A. Thitithammawong, J. W. Noordermeer, A. Kaesaman and C. Nakason. Journal of Applied Polymer Science, 2008, 2436-2443.
[5] T. Prasopdee, D. U. Shah, and W. Smitthipong. Macromolecular Materials and Engineering, 2021, 202100337.

IOP Conf. Series: Materials Science and Engineering **1325** (2025) 012008 doi:10.1088/1757-899X/1325/1/012008

# QM/MM/3D-RISM study of an electronic structure change of benzimidazole derivative induced by molecular recognition of cucurbit[7]uril

**Kodai Kanemaru,[12] Keisuke Shiroshita,[3] Haruyuki Nakano, [3] Norio Yoshida[2*]**

[1] Fukui Institute for Fundamental Chemistry, Kyoto University, Takano-Nishibiraki-cho 34-4, Sakyo-ku, Kyoto 606-8103, Japan
[2] Department of Complex Systems Science, Graduate School of Informatics, Nagoya University, Furocho, Chikusa-ku, Nagoya 464-8601, Japan
[3] Department of Chemistry, Graduate School of Science, Kyushu University, Motooka 735, Nishi-ku, Fukuoka 819-0395, Japan

*Corresponding author: noriwo@nagoya-u.jp

**Abstract.** The quantum mechanics/molecular mechanics three-dimensional reference interaction site model (QM/MM/3D-RISM) method is a three-layer hybrid approach designed to calculate the electronic structure of complex molecules in solution. In this study, the QM/MM/3D-RISM method was employed to investigate electronic structure changes in ligand molecules upon molecular recognition by the macrocyclic host molecule cucurbit[7]uril (CB7). Thiabendazole (TBZ), a benzimidazole derivative, was selected as the ligand, which is known for its significant alterations in properties such as p$K_a$, solubility, photostability, and fluorescence intensity upon binding to CB7. The calculations elucidated the mechanism underlying the changes in the electronic structure induced by encapsulation, as well as the role of solvation in these transformations.

## 1. Introduction

Molecular recognition (MR) is a process where ligand molecules bind to specific sites on host molecules, such as proteins, inducing changes in molecular properties. This makes MR a valuable tool for tuning molecular functions. Among various host molecules, macrocyclic compounds like cucurbiturils (CBs) are particularly advantageous due to their lower toxicity, higher binding affinities, and easier synthesis compared to cyclodextrins. Structurally, CBs are oligomers of glycoluril monomers linked by methylene bridges, making them promising candidates for applications like drug delivery and self-healing materials. Benzimidazole derivatives, widely used in pharmaceuticals, show altered physicochemical properties, including p$K_a$, solubility, and fluorescence, upon encapsulation in CB7 [1].

Theoretical investigations of molecular properties in encapsulated systems require a methodology capable of addressing several challenges. First, since the changes in properties are fundamentally linked to alterations in the electronic structure of the ligand molecule, quantum chemical methods are essential. Second, the size of the macromolecular systems, such as CB7, necessitates an approach that can efficiently handle large molecules. Finally, given that MR occurs in solution, an accurate treatment of solvent effects is crucial.

IOP Conf. Series: Materials Science and Engineering **1325** (2025) 012008 doi:10.1088/1757-899X/1325/1/012008

In this study, the quantum mechanics/molecular mechanics three-dimensional reference interaction-site model (QM/MM/3D-RISM) method—a three-layered hybrid approach—was employed to address these requirements [2]. The QM/MM/3D-RISM method partitions the system into three regions: the chemically active QM region, the inactive MM model region, and the solvent which described by the 3D-RISM theory. In this study, we investigated the electronic structure changes in thiabendazole (TBZ) induced by MR with CB7 in aqueous solution. TBZ, as the ligand, was treated with the QM method, CB7 as the host is assigned to the MM region, and the water solvent was modeled using the 3D-RISM theory.

## 2. Theory

A brief explanation of the QM/MM/3D-RISM method is given here, as the details can be found in original paper.[2] In the QM/MM/3D-RISM formalism, the system consists of solute ligand and host molecules immersed in solvent at infinite dilution. The ligand electronic Hamiltonian is given by

$$\hat{H}(\boldsymbol{x},\boldsymbol{X}) = \hat{H}_0(\boldsymbol{x}) + \hat{V}_{\mathrm{MM}}(\boldsymbol{X}) + \hat{V}_{\mathrm{3DRISM}}(\boldsymbol{x},\boldsymbol{X})$$

where $\hat{H}_0(\boldsymbol{x})$ is the electronic Hamiltonian of ligand with the atomic coordinate $\boldsymbol{x}$ in isolated system. The potential acting on the ligand electron due to the host molecule $\hat{V}_{\mathrm{MM}}(\boldsymbol{X})$ and the potential acting on the ligand electron by the solvent molecules $\hat{V}_{\mathrm{3DRISM}}(\boldsymbol{x},\boldsymbol{X})$ are given as:

$$\hat{V}_{\mathrm{MM}}(\boldsymbol{X}) = -\sum_{i}^{\mathrm{MM}} \frac{q_i}{|\boldsymbol{X}_i - \boldsymbol{r}_e|}, \quad \hat{V}_{\mathrm{3DRISM}}(\boldsymbol{x},\boldsymbol{X}) = -\sum_{j}^{\mathrm{solvent}} \rho_j \int \frac{q_j^v}{|\boldsymbol{r} - \boldsymbol{r}_e|} g_j(\boldsymbol{r};\boldsymbol{x},\boldsymbol{X}) d\boldsymbol{r}$$

where $\boldsymbol{X}_i$ denotes the coordinate of MM atom $i$ and $\boldsymbol{X}$ is the set of $\boldsymbol{X}_i$. $q_i$ is the point charge on the MM atom $i$ and $\boldsymbol{r}_e$ is the coordinate of electron. $\rho_j$ and $q_j^v$ are the number density and point charge of solvent site $j$. $g_j(\boldsymbol{r};\boldsymbol{x},\boldsymbol{X})$is the distribution function of solvent site $v$ which is parametrically depends on QM and MM coordinates, $\boldsymbol{x},\boldsymbol{X}$. The distribution function should be obtained by solving the 3D-RISM equation coupled with the closure relation such as Kovalenko-Hirata closure with the solute-solvent interaction potential which depends on the solute electronic wave functions.

## 3. Computational details

### 3.1 Parameter preparation

In this study, we employed the QM/MM/3D-RISM method to evaluate the electronic structure of TBZ in solution. It is known that the TBZ takes protonated state, $TBZH^+$, depending pH. Therefore, in this study, we consider both TBZ and $TBZH^+$ as ligands. The initial structure of TBZ, $TBZH^+$, and CB7 were determined by the geometry optimization using Gaussian 16 with B3LYP/6-31G(d) level calculations.[4] The antechamber in AmberTools20 were employed for the parameter assignment.[5] The GAFF parameter set were used for these molecules and the RESP method was used to determine the point charges on solute atoms.[6] The TIP3P forcefield was employed for the solvent water molecules.[7]

### 3.2 Structure sampling

In prior to the QM/MM/3D-RISM calculation, the QM/MM simulations with explicit TIP3P water were conducted to structure sampling. The PM3 empirical quantum chemical method was applied for ligand TBZ and $TBZH^+$. The 4 ns production NPT MD simulations were conducted for each system with Berendsen temperature and pressure control at 300 K and 1 bar following 20 ps equilibration. The SHAKE method was applied to H atoms and 2 fs time step is used. A cubic solvation box filled with water molecules up to a distance of 15 Å from the solute molecules was prepared. The number of water molecules was about 1700 and 2600 for free ligands and CB7-ligand complex systems, respectively.

### 3.3 Electronic structure and solvation structure calculation

QM/MM/3D-RISM was performed for the solute structure extracted from the QM/MM simulation and striped all the explicit water at every 10 ps for 4 ns. Thus, 400 structures were used in QM/MM/3D-RISM. The B3LYP/6-31G(d) level was employed for the QM calculation. The density of solvent water was set to 0.99 g/cm$^3$ and 64 Å × 64 Å × 64 Å solvent box with 0.5 Å grid spacing was employed for the 3D-RISM calculations. The Kovalenko-Hirata closure was used. All the QM/MM/3D-RISM calculations were conducted the RISMiCal program package combined with GAMESS quantum chemical program package.[8-11]

## 4. Results and discussion

### 4.1 Complex structure

Figure 1(a) and (b) depict the superimposed complex structures of TBZ-CB7 and TBZH$^+$-CB7. As shown, both TBZ and TBZH$^+$ are bound to CB7 through their benzimidazole groups, which is consistent with observations from NMR experiments [1]. Specifically, the nitrogen atoms in the benzimidazole ring form hydrogen bonds with the carbonyl oxygen atoms of CB7, highlighting the key interactions responsible for the binding.

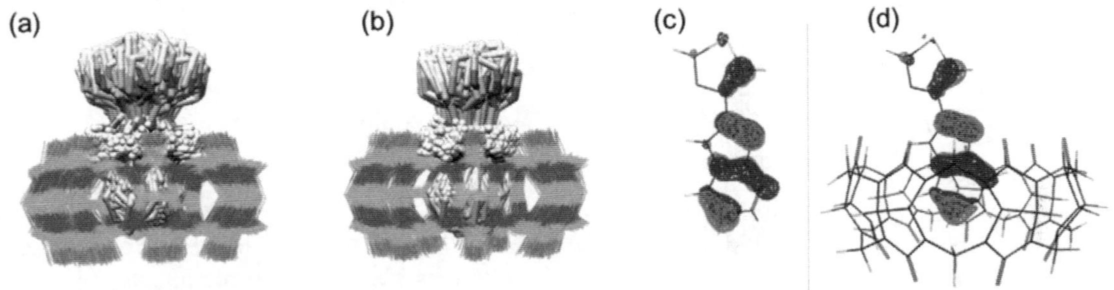


**Figure 1.** Complex structure of (a)TBZ and (b)TBZH$^+$ with the CB7, respectively. HOMO of TBZ in (c) free and (d) complex form, respectively.

### 4.2 Electronic structure

In this subsection, we discuss the electronic structure changes resulting from complex formation. Figure 1(c) and (d) compare the highest-occupied molecular orbital (HOMO) of TBZ in its free state and complexed with CB7, based on one of the snapshot structures. The HOMO primarily consists of π-bonds in the aromatic rings. The encapsulation by CB7 appears to have minimal impact on the shape of the HOMO.

Table 1 summarizes the averaged dipole moments and HOMO-LUMO gaps of TBZ and TBZH$^+$ in their free and complex states. For TBZ, the dipole moment undergoes a more significant change upon encapsulation compared to TBZH$^+$. Specifically, the dipole moment of TBZ increases by approximately 1 Debye due to encapsulation. In contrast, the HOMO-LUMO gap is less affected by encapsulation and more influenced by changes in the protonation state. These findings suggest that the changes in the absorption spectrum are likely driven by protonation effects rather than direct interactions with CB7 during encapsulation.

**Table 1.** The dipole moment and HOMO-LUMO gap of free and complex ligand

| | TBZ | | TBZH$^+$ | |
|---|---|---|---|---|
| | Free | Complex | Free | Complex |
| Dipole moment / Debye | 4.99 | 5.96 | 2.30 | 2.23 |
| HOMO-LUMO gap/ eV | 4.58 | 4.56 | 4.37 | 4.27 |

### 4.3 Solvation structure

In addition to the interaction with CB7, changes in the solvation structure may significantly influence the electronic structure of TBZ. Figure 2 illustrates the spatial distribution of water oxygen and hydrogen, represented by red- and gray-colored surfaces, respectively. For free TBZ, a prominent peak of hydrogen-bonded solvent oxygen is observed around the imide hydrogen of the benzimidazole group. Upon binding to CB7, this solvent oxygen is replaced by the carbonyl oxygen of CB7. A distinct peak of solvent hydrogen is also observed near the unprotonated imide group. Interestingly, this hydrogen

IOP Conf. Series: Materials Science and Engineering **1325** (2025) 012008 doi:10.1088/1757-899X/1325/1/012008

peak persists in the complex state, forming a bridge between the nitrogen atom of TBZ and the carbonyl oxygen of CB7. Furthermore, the hydrophobic aromatic rings of TBZ are encapsulated within CB7 and are partially dehydrated compared to their free state. For $TBZH^+$, both imide hydrogens exhibit peaks of hydrogen-bonded water oxygen in the free state, but these peaks disappear upon binding to CB7.

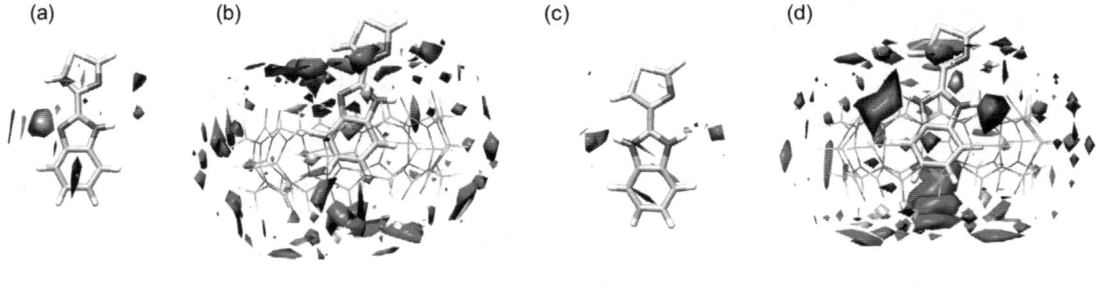


**Figure 2.** 3D distribution of solvent oxygen (in red, $g(\boldsymbol{r}) = 2.8$) and hydrogen (in gray, $g(\boldsymbol{r}) = 2.3$). (a) Free TBZ, (b) complex, CB7-TBZ, (c) free $TBZH^+$ and (d) complex CB7-$TBZ^+$

## 5. Conclusions

In this study, the electronic and solvation structures of thiabendazole (TBZ) and its protonated form ($TBZH^+$) in their free states and in complex with cucurbit[7]uril (CB7) are theoretically investigated using the QM/MM/3D-RISM method. While the results presented here are preliminary, they offer valuable insights into the changes in TBZ properties associated with molecular recognition. Future work will include a more comprehensive evaluation of computational parameters, such as functionals and basis sets, to enable a more detailed investigation.


## Acknowledgement

This work was supported by MEXT as "Program for Promoting Researches on the Supercomputer Fugaku"(JPMXP1020230327) and used computational resources of supercomputer Fugaku (hp230212, hp240223) and the MEXT Program: Data Creation and Utilization-Type Material Research and Development Project Grant No. JPMXP1122714694. The computations were also performed using the Research Center for Computational Science, Institute for Molecular Science (23-IMS-C068 and 24-IMS-C066) and MCRP-S at the Center for Computational Sciences, University of Tsukuba. This work was also supported by the JSPS KAKENHI (22H05089, 24K01434, 22H01873, 24H00282) and the Toyota Physical and Chemical Research Institute scholar collaborative research project.


## References


[1] Koner A L, Ghosh I, Saleh N and Nau W M 2011 *Can. J. Chem.* **89** 139
[2] Yoshida N, Kiyota Y and Hirata F 2011 *J. Mol. Liquids* **159** 83
[3] Kovalenko A and Hirata F 1999 *J. Chem. Phys.* **110** 10095
[4] Frisch M J et al. Gaussian 16, Rev. C.01, Gaussian, Inc., Wallingford CT, 2019
[5] Case D A et al., Amber 2020
[6] Wang J, Wolf R M, Caldwell J W, Kollman P A and Case D A, 2004 *J. Comput. Chem.* **25** 1157
[7] Jorgensen W L, Chandrasekhar J, Madura J D, Impey R W and Klein M L 1983 *J. Chem. Phys.* **79** 926
[8] Yoshida N and Hirata F 2006 *J. Comput. Chem.* **27** 453
[9] Yoshida N 2014 *J. Chem. Phys.* **140** 214118
[10] Maruyama Y and Yoshida N 2024 *J. Comput. Chem.* **45** 1470
[11] Barca G M J et al. 2020 *J. Chem. Phys.* **152** 154102

# Enhance fire behavior of rigid polyurethane foam by intumescent flame retardants

**V Peasuwan[1], W Keawwattana[1*], H Paeprapa[1], M Raoarun[1] and K Hancharoen[2]**

[1] Department of Chemistry, Faculty of Science, Kasetsart University, Bangkok 10900, Thailand
[2] Department of Architecture, Faculty of Architecture, Kasetsart University, Bangkok 10900, Thailand

*Corresponding author: fsciwyk@ku.ac.th

**Abstract.** Rigid polyurethane foam (RPUF) is the most widely used. Nonetheless, it is high ignitability. In order to guarantee safety, the incorporation of flame retardant is required. In this work, the improvement of flame retardant performance of RPUF is studied using layer by layer (LbL) techniques, containing intumescent flame retardant (IFR): Ammonium dihydrogen phosphate, Neopentyl glycol and Melamine in the ratio of 2:1:1 wt%. For the optimum coating condition, it was found that the ratio of polyethyleneimine and deionized water 1:2 wt% provided the suitable viscosity solution for coating and takes less drying time 30 min at 150 °C and 75 wt% of IFR capable of improving the flame retardant properties of RPUF. The result of ignitability test revealed that 75 wt% of IFR can effectively coat on RPUF and substantially improve its flame retardant properties by suppressing the melt dripping during flammability and immediately extinguishing flame after removing the ignition

## 1. Introduction

Polymer based products have been widely used in our life for their cheap, light weight and easy to be processed. Rigid polyurethane foam (RPUF) is one of polymer and the most widely used interior materials due to its excellent resilience and mechanical property. Nonetheless, it has high fire vulnerability due to its high ignitability and produces a lot of harmful smoke. In order to guarantee the safety when they exposed to fire risk, the flame retardants (FR) were used, thus significantly enhancing fire safety of materials [1]. The widely used flame retardant additives are intumescent flame retardants. The flame retardant mechanism usually involves reducing the temperature via heat absorption, reducing the content of oxygen, forming a protective film to isolate the material from oxygen, and capturing free radicals to inhibit the combustion process [2]. Intumescent coatings are composed of three flame retardant additives: an acid source (such as ammonium polyphosphate and APP), a carbon source (such as pentaerythritol and PER), and a blowing agent (such as melamine and MEL). The benefits of flame retardants are form intumescence char layer when exposed to heat, which acts as a thermal barrier that effectively protects the substrate against fire, thereby maintaining the structural of material and the during combustion has low smoke and low toxicity but the flame retardants have drawbacks as reducing flexibility, strength, or durability of material and the incorporation of flame retardants into products may increase manufacturing costs due to the use of specialized chemicals or more complex production processes [3].

Recently, layer by layer (LbL) assembly is simple and has been described as a good candidate for satisfying the requirement for green flame retardants to protect the substrate and prefer the char formation. Lately the coatings have been proposed to represent intumescent systems by integrating an acidic source, a carbon source, and a blowing agent within the assembly. The main benefits of coating do not compromise the internal structure of the material and preserves properties [4]. In this work, we aim to study the LbL assembly of an efficient FR coating comprising Intumescent flame retardants, Ammonium dihydrogen phosphate (ADP), Pentaerythritol (PER) and Melamine (MEL) for the protection of open cell RPUF. The main aim is the production of an efficient LbL coating capable of delivering strong FR performances with a reduced number of deposition cycles to reduce the use of chemicals. The growth of IFR assembly was monitored with FT-IR spectroscopy. Flammability behaviour of untreated and treated foams have been investigated by the ignitability test with single-flame source (EN ISO 11925-2).

## 2. Experimental

### 2.1 Materials

Polyurethane foam (PU) was purchased from Bangkok Panel Insulation Co.,Ltd. The primary components of this investigation were ammonium dihydrogen phosphate (ADP), Melamine (MEL), Pentaerythritol (NPG) and Polyethyleneimine (PEI) were supplied by Apex Chemical Co.,Ltd.

### 2.2 Preparation of flame retardant solutions

The flame retardant solution was prepared by varying the ratio of PEI and DI water to receive the appropriate viscosity as displayed in Table 1.

**Table 1** Ratio of PEI and DI water

| No. | Ratio of IFR (100 wt%) | | | Ratio of PEI: DI water | |
|---|---|---|---|---|---|
| | **ADP** | **NPG** | **MEL** | **PEI** | **DI water** |
| 1 | 2 | 1 | 1 | 3 | 0 |
| 2 | | | | 1 | 1 |
| 3 | | | | 1 | 2 |
| 4 | | | | 2 | 1 |

### 2.3 Layer by layer deposition

After obtaining the viscosity ratio of the solution from Table 1, the IFR solutions with varying content (50%, 75% and 100%) as shown in Table 2 were coated on the RPUF using a spatula to spread the solution evenly on all sides. LbL-treated foams were dried to constant weight in a ventilated oven at 150 °C for 30 min. The process was repeated until 4 layers were built on PU foam. The differences were observed and recorded.

**Table 2** The various concentrations of intumescent flame retardant (IFR)

| No. | Ratio of PEI: DI water | | Ratio of IFR | | | Intumescent (wt%) |
|---|---|---|---|---|---|---|
| | **PEI** | **DI water** | **ADP** | **NPG** | **MEL** | |
| 1 | 1 | 2 | 2 | 1 | 1 | 50 |
| 2 | | | | | | 75 |
| 3 | | | | | | 100 |

IOP Conf. Series: Materials Science and Engineering **1325** (2025) 012009 doi:10.1088/1757-899X/1325/1/012009

2.4 Characterization

Flammability behaviour of untreated and treated foams have been investigated by the ignitability test with single-flame source to determine the ignitability of a vertically oriented test specimen when exposed to a small flame (EN ISO 11925-2). The growth of IFR assembly was monitored with Fourier Transform Infrared Spectroscopy (FTIR).

## 3. Results and discussion

3.1 Preparation of flame retardant solutions.

**Table 3** Ratio of PEI and DI water

| **Ratio of PEI: DI water** | | | |
|---|---|---|---|
| **Sample 1** | **Sample 2** | **Sample 3** | **Sample 4** |
| **3:0** | **1:1** | **1:2** | **2:1** |

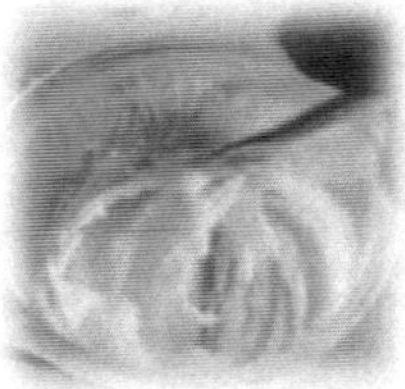
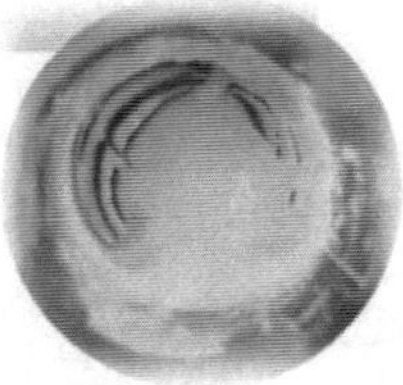
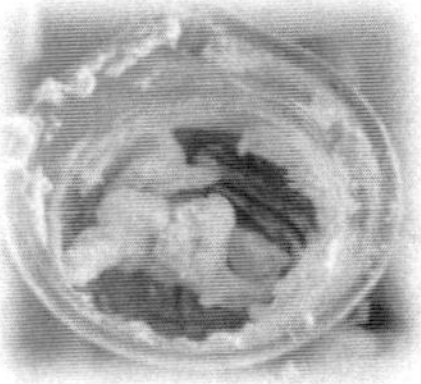

The experiment revealed that when using only PEI as a bonding agent, the resulting substance exhibited high stickiness and viscosity, which led to slow drying and difficulty in achieving an even coating. When DI water was added to dilute the viscosity in a 1:1 ratio, the excess liquid solution tended to drip when applied to RPUF. However, for Sample 3, where the ratio of PEI to DI water was 1:2 wt%, the viscosity remained manageable and allowed for an even coating on RPUF. The fastest drying time observed was 30 min at a temperature of 150°C.

3.2 Layer by layer techniques on polyurethane foam

After determining the optimal viscosity ratio of the solution, the coating was applied to the RPUF. The formulation that yielded the best results consisted of an intumescent material with a ratio of ADP: NPG: MEL at 2:1:1 wt%, combined with a PEI: DI water ratio of 1:2wt%. This combination allowed for an even and complete coating on the surface of the RPUF. To evaluate the flame retardancy of the intumescent material, an ignitability test was conducted using the single-flame source test. The test was performed positioned at a 90° angle, at the center of the flame. Ignition behavior was observed at two intervals: 15 s and after 30 s, as detailed in Table 4.

**Table 4** The result of ignitability test with single-flame source (EN ISO 11925-2)

| **No.** | **Ratio of PEI: DI water** | | **Coating (cycle)** | **The result of Ignitability test** | |
|---|---|---|---|---|---|
| | **PEI** | **DI water** | | **15 s** | **30 s** |
| RPUF | | | 0 | Positive | Positive |
| 1 | | | 1 | Negative | Negative |
| 2 | 1 | 2 | 2 | Negative | Negative |
| 3 | | | 3 | Negative | Negative |
| 4 | | | 4 | Negative | Negative |

IOP Conf. Series: Materials Science and Engineering **1325** (2025) 012009 doi:10.1088/1757-899X/1325/1/012009

The test results were negative for both the first 15 s and the second 30 s, indicating that the intumescent coating effectively controlled the ignition and prevented the spread of fire to RPUF and found that a single coating application was sufficient to delay the ignition of the RPUF.

### 3.3 Effect of IFR content on coating efficiency and flammability test

The experiment on reducing the concentration of the intumescent material revealed that at a 50% concentration, the solution exhibited lower viscosity, making it more difficult to apply and causing dripping during the coating process. At a 75% concentration, however, the solution had an optimal viscosity, allowing for a complete and even coating on the RPUF. This coating did not drip during either the application or baking processes and adhered fully to the surface of the RPUF, demonstrating performance comparable to that of the 100% intumescent concentration. Furthermore, a single coat of the 75% intumescent solution on the RPUF effectively retarded fire after 15 s and prevented ignition at 30 s, similar to the results achieved with the 100% concentration. The combustion characteristics are summarized in Table 5.

**Table 5** The characteristics of coating IFR (before and after) burning

| **Ratio of chemical (wt%)** | **PEI: DI water (1:2)** | | |
|---|---|---|---|
| | **IFR (ADP: NPG: MEL) wt% (2:1:1)** | | |
| **IFR (wt%)** | **50** | **75** | **100** |
| **Coated 1 time** | 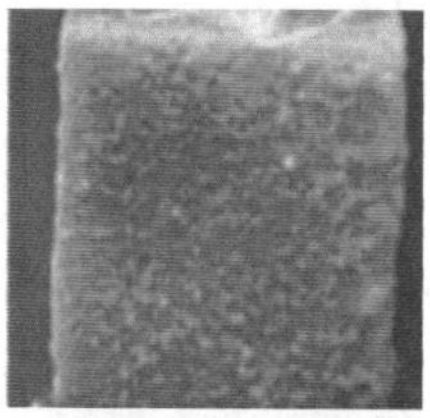 |  |  |
| **The ignitability test** | 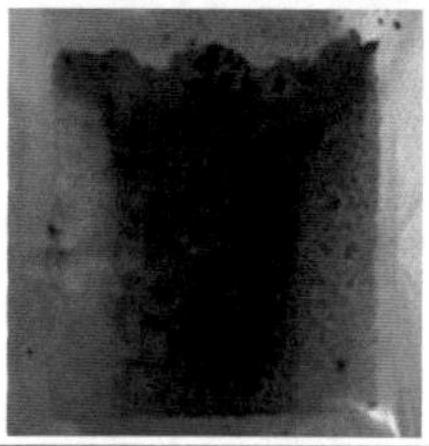 |  | 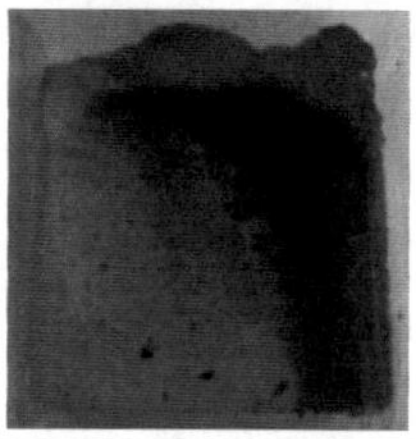 |

### 3.4 Coating growth by FT-IR spectroscopy

The LbL growth before (a) and after (b) combustion was monitored using FT-IR spectroscopy as seen in Figure 1. It was found that the peaks of various structural components after combustion shifted from their original positions, the peak at 2385 $cm^{-1}$, which is attributed to the decomposition of ADP, diminished over time. This indicates the release of $NH_3$. As combustion continued, the intensity of the peaks decreased, suggesting structural deformation. At the peak of 1553 $cm^{-1}$, Triazine ring bonding in the MEL structure [5], became evident. This peak signifies the occurrence of chemical reactions at various stages in the flame retardancy process of the intumescent system. These reactions contribute to the formation of a charred layer, which plays a crucial role in preventing the spread of fire in rigid polyurethane foam.

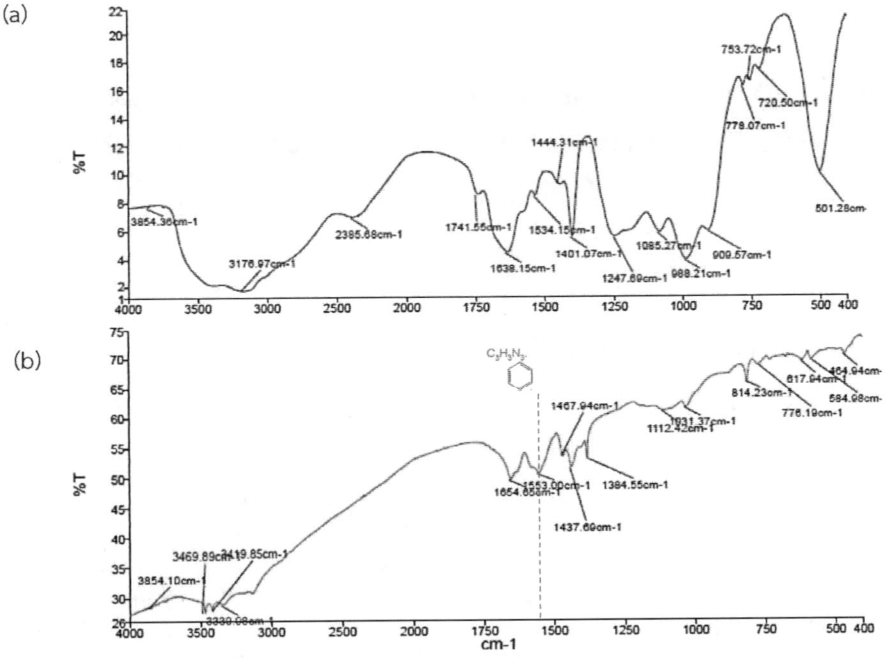


**Figure 1**. FT-IR spectra of IFR before (a) and after (b) burning

## 4. Conclusions

In this research, show multilayered structures comprised by IFR have been assembled using LbL technique and employed as flame retardant coating on PU foams. The flammability test revealed that the intumescent coating effectively controlled the ignition and prevented the spread of fire to RPUF and the ratio of 75% IFR (ADP: NPG: MEL at 2:1:1 wt%) combined with a PEI: DI water ratio of 1:2 wt% and the coating a single coating application was sufficient to delay the ignition of the RPUF.

## Acknowledgement

The research was financially supported by the Department of Chemistry, Faculty of Science, Kasetsart University. The authors are grateful to the Specialized Center of Rubber and Polymer Materials for agricultural and industry (RPM), Faculty of Science, Kasetsart University for presentation support.

## References

[1] Choi W, Kim W, Kwon S, Kang W, Song I and Park T 2021 *J. Polym.* 13.
[2] Wang Y, Liu J, Pan X, Zhao M and Zhang J 2023 *J. Polym.* 15.
[3] Yew M-C and Ramli Sulong H 2012 *Mater. Design* **34** 719-724.
[4] Carosio F, Negrell-Guirao C, Alongi J, David G and Camino G 2015 *J. European Polym.* **70** 94-103.
[5] Wang W, Wen P, Zhan J, Hong N, Cai W, Gui Z and Hu Y 2017 *Polym. Degradation and Stability* **144** 454-463.

# Findings the proper high styrene resin/accelerator/antioxidant ratios to natural rubber/reclaimed rubber composites: Cure characteristics, mechanical properties and weathering resistance

**P Boochatum[1], W Mingvanish[1] and Y Nakaramontri[1*]**

[1]Sustainable Polymer and Innovative Composite Materials Research Group, Department of Chemistry, Faculty of Science, King Mongkut's University of Technology, Thonburi, Bangkok 10140, Thailand
*Corresponding author: Yeampon Nakaramontri (yeampon.nak@kmutt.ac.th)

**Abstract.** This study presents the effects of varying ratios of high styrene rubber (HSR), carbon black (CB), N,N'-dibenzylthiuram disulfide (MBTS), and 2,2,4-trimethyl-1,2-dihydroquinoline (TMQ) on the cure characteristics, mechanical properties, and aging stability of natural rubber (NR)/reclaim rubber (RR) composites. The addition of TMQ increased cure time and torque values but reduced tensile properties due to defects in the rubber matrix. Increasing HSR content reduced crosslink density, while higher MBTS concentrations improved crosslinking and reduced cure time. NR/RR composites with optimized ratios (10:60:3:0-2 and 10:60:4:0-2) exhibited the highest tensile strength and elongation at break. Weathering tests showed that UV exposure improved tensile strength and modulus, particularly with higher MBTS concentrations, enhancing chemical stability. These findings suggest that HSR, TMQ, CB, and MBTS ratios can optimize NR/RR composites for improved durability in harsh conditions for applying in the formulation of outdoor applications.

## 1. Introduction

Rubber composites are widely utilized in various industrial applications relating to their versatility, durability, and excellent mechanical properties [1]. Among these composites, reclaim rubber (RR) has garnered significant attention as an eco-friendly alternative to virgin rubber, offering a cost-effective solution for reducing the environmental impact of waste rubber [2]. However, the performance of RR composites can be limited by factors such as poor crosslinking, thermal instability, and weathering resistance. Therefore, optimizing the composition and curing systems of RR composites is crucial for enhancing their mechanical and functional properties. This can be solved by combination of RR with fresh rubber, in particular the natural rubber (NR) which contains the highest molecular weigth and long mainchain rubber relative to other synthetic rubber [3]. However, combination of RR and NR needs the chemical crosslinking among rubber molecules for improving properties together with the effective formulation designs.

Chemical crosslinking causes a vital role in improving the mechanical properties and durability of rubber composites. Crosslinking agents such as sulfur are commonly used, but their effectiveness can be influenced by the type and concentration of accelerators and other additives. The incorporation of fillers like carbon black (CB) further enhances the reinforcement, while chemical activators such as

MBTS (N,N'-Dibenzylthiuram disulfide) improve crosslinking efficiency [4]. In addition, antioxidants such as TMQ (2,2,4-Trimethyl-1,2-dihydroquinoline) are often added to enhance the thermo-oxidative stability of rubber composites, preventing premature degradation under harsh environmental conditions [6].

Recently, significant efforts have been made to optimize the formulation of NR/RR composites by adjusting the ratios of HSR, CB, MBTS, TMQ, and other chemicals inside the composites [7,8]. These modifications are intended to improve the chemical crosslinking characteristics, tensile properties, aging stability, and weathering resistance of the composites. Cure characteristics provide valuable insights into the crosslinking behavior, while tensile properties such as tensile strength, elongation at break, tear strength, and modulus offer an understanding of the composite's performance. Furthermore, aging and weathering tests are essential to assess the long-term stability of these materials under real-world environmental conditions, including UV exposure and high temperatures [9,10]. Thus, with the need to improve the durability of rubber products, the performance of HSR is influenced by its concentration and the proper ratio with accelerators and antioxidants. This study aims to elucidate and examine these factors to suggest the optimal concentration of HSR, accelerator, and antioxidant for enhancing the performance of rubber products, particularly in outdoor applications requiring weather resistance. This remains a promising area of research with no prior definitive determination.

The present work aims to examine the effects of varying the ratios of HSR, CB, MBTS, and TMQ on the cure characteristics, mechanical properties, and aging stability of NR/RR composites. Through a series of experimental tests, including tensile testing, weathering resistance (QUV), and aging studies, we seek to optimize the formulation for improved durability and performance in challenging environments. The findings of the work will lead to the development of high-performance NR/RR composites that can be applied in a wide range of industries, including automotive, construction, and manufacturing, while also promoting sustainable practices through the use of reclaimed materials.

## 2. Experimental

### 2.1 Materials

Natural rubber (NR), reclaim rubber (RR) and high styrene resin (HSR) were supported by Thai Rubb Tech Co., Ltd. (Samut Sakhon, Thailand). Carbon black (CB), Vulcan XC72, with 30 nm diameter was purchased by Cabot Corporation (Pampa, TX, USA). Other chemicals used in the NR formulation were activator, accelerator and sulfur curing agent were purchased from different companies, as can be summarized in Table 1.

**Table 1** Compounding formulation of NR composites and their chemical suppliers.

| Chemicals | Suppliers | Contents (phr*) |
|---|---|---|
| NR/RR | Thai Rubb Tech Co., Ltd. | 100 |
| HSR | Thai Rubb Tech Co., Ltd. | 0-20 |
| CB | Cabot Corporation Co., Ltd. | 60-80 |
| MBTS | Imperial Chemical Co., Ltd. (Pathum Thani, Thailand) | 2-4 |
| TMQ antidegradant | Flexsys Inc., (Termoli, Italy) | 0-2 |
| Sulfur | Ajax Chemical Co., Ltd. (Samut Prakarn, Thailand) | 1 |

* phr = part per hundred rubber

### 2.2 Preparation of NR/RR compounds filled with different ratios of HSR:CB:MBTS:TMQ

The NR/RR compounds filled with various HSR:CB:MBTS:TMQ loadings were carried out in an internal mixer (Model MX75; Charoen tut Co., Ltd., Samutprakarn, Thailand) at 80°C, 60 rpm and filled factor of 75%. The NR was initially masticated for decreasing molecular weight and viscosity for 3 min followed the condition of other ingredients and conditioned mixing in different duration times by CB and HSR for 6 min, activators and plasticizer for 2 min, accelerator and antidegradants for 2 min and finally the sulfur curing agent for another 2 min. The compounds were then sheeted with a two-roll mill for improving the dispersion and distribution of an additional additive and fillers.

### 2.3 Characterization

#### 2.3.1 Cure characteristics

Chemical crosslinking characteristics of the NR/RR compounds were examined using a moving die rheometer (MDR) (Monsanto Co., Ltd, Ohio, USA). A fixed oscillation frequency of 1.66 Hz with 1 arc degree amplitude were applied under the temperature of 160°C.

#### 2.3.2 Mechanical properties

The dumb-bell shaped specimens are prepared following ISO 527 Type 2and performed the tests through a Zwick Z 1545 Universal testing machine (Zwick GmbH & Co. KG, Ulm, Germany) under an ambient condition (25 ± 2°C). The 100%modulus, tensile strength, tear strength and elongation at break of the vulcanizates were reported under the extension speed of 500 mm/min through 9 reciprocal specimens. In addition, hardness of the NR/RR vulcanizates was tested using the Shore A hardness tester regarding to ASTM D2240-05.

#### 2.3.3 Thermal aging

Thermal stability of the vulcanizates was measured from the dumb-bell shaped specimens which were conditioned in an air circulating oven at 100°C and 150°C for 72 h and 24 h, respectively. After that, the specimens were kept at room temperature of at least 18 h before the tests were started. The tensile properties, including the 100%modulus, tensile strength, and elongation at break were then elucidated through the same protocol as the unconditioned specimens. Then, the relative properties were calculated using Equation (1) for interpretated as the thermal ageing.

$$\text{Property retention value} = \frac{P}{P_0} \times 100 \tag{1}$$

where $P_o$ and $P$ are the properties of the specimens with and without conditioning regarding to the tensile strength and elongation at break, respectively.

#### 2.3.4 Weathering resistance

The accelerated aging was investigated through the ASTM G154 cycle I using the QUV-accelerated weathering tester (Q-Lab Corporation, Westlake, USA). Testing condition was set at 60 h, including the applied UV-A radiation of wavelength for 340 nm and intensity of 0.77 W/m$^2$ at 60°C for 8 h and following with the water-condensation condition at 50°C for 4 h continuing with water spraying to the specimen surface within 15 min. In addition, the specimens were through the ASTM G154 cycle VII using the same equipment as the condition of an applied UV-B radiation under the wavelength of 340 nm, intensity of 1.55 W/m$^2$ with temperature of 60°C for 8 h before doing the water spraying for another 15 min together with water-condensation at 50°C for 4 h. So that, the total weathering testing is also 60 h. After conditioning at room temperature for another 18 h over the applied environmental damaging, the specimens were tested the properties using the Universal tensile testing following the ASTM D 412-06 through the five reciprocal samples with an extension speed of 500 mm/min.

## 3. Results and discussion

### 3.1 Cure characteristics

Figure 1 illustrates the chemical crosslinking behaviors in NR/RR composites with varying HSR/CB/MBTS/TMQ ratios, while Table 2 summarizes the minimum torque ($M_L$), maximum torque ($M_H$), torque difference ($M_H$-$M_L$), scorch time ($T_{s2}$), and cure time ($T_{90}$) for each specimen. The data indicate that the addition of TMQ slightly increases $M_H$-$M_L$, $T_{s2}$, and $T_{90}$, as observed under the plateau cure curves. The increase in $M_H$-$M_L$ with the addition of TMQ is attributed to the rise in $M_H$, indicating improved thermo-oxidative resistance of the vulcanizates. Meanwhile, the longer $T_{s2}$ and $T_{90}$ are related to enhanced thermal resistance, which slightly delayed thermal extension throughout the compounds. In addtion, it is seen that the HSR content is increased to 20 phr while reducing the CB loading, a decrease in the estimated crosslink density and cure time is observed. This is attributed to the reduced reinforcement efficiency and enhanced thermal expansion due to the high thermal conductivity of CB. For formulations with reduced CB:MBTS ratios, specifically at 60:3 and 60:4 phr, shorter cure times are observed along with an increase in crosslink density. This is due to the higher MBTS concentration, which accelerates chemical crosslinking based on sulfur curing mechanisms. In summary, the curability of the RR composites is slightly enhanced with TMQ incorporation, while the crosslink propagation benefits from the synergistic effects of thermally conductive CB and increased MBTS concentrations.

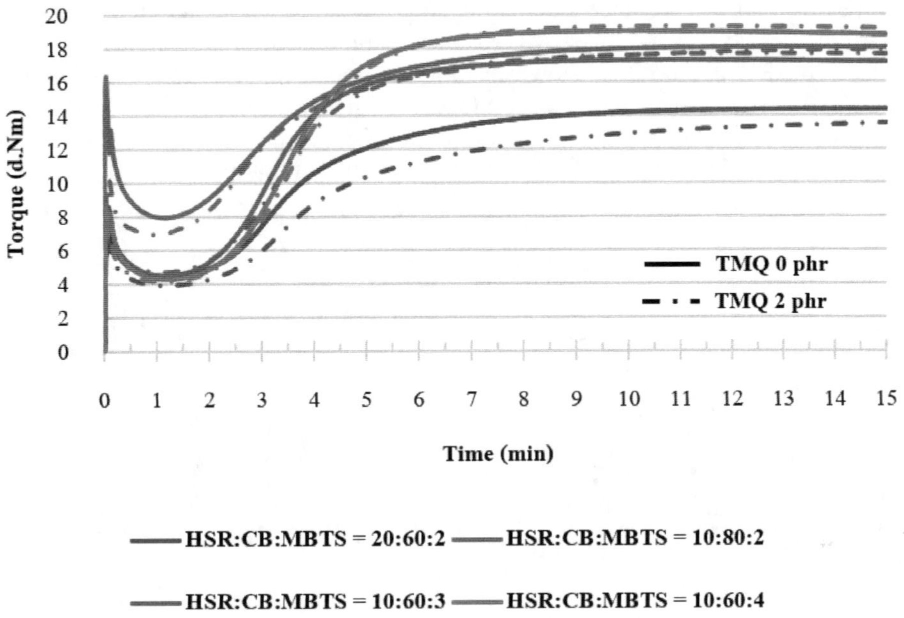


**Figure 1** Cure characteristics curves of the NR/RR composites filled with different ratios of HSR/CB/MBTS/TMQ

**Table 2** Summarization of the $M_L$ and $M_H$, $M_H$-$M_L$ together with $T_{s1}$ and $T_{90}$ of NR/RR composites filled with different ratios of HSR/CB/MBTS/TMQ.

| HSR:CB:MBTS:TMQ (phr) | $M_L$ (d.Nm) | $M_H$ (d.Nm) | $M_H$-$M_L$ (d.Nm) | $T_{S2}$ (min) | $T_{90}$ (min) |
|---|---|---|---|---|---|
| 20:60:2:0 | 4.41 | 14.36 | 9.95 | 02:42 | 06:47 |
| 20:60:2:2 | 3.91 | 13.51 | 9.60 | 02:58 | 08:37 |
| 10:80:2:0 | 7.96 | 18.08 | 10.12 | 02:17 | 06:14 |
| 10:80:2:2 | 6.93 | 17.91 | 10.98 | 02:19 | 07:00 |
| 10:60:3:0 | 4.49 | 17.30 | 12.81 | 02:24 | 05:11 |
| 10:60:3:2 | 4.71 | 17.87 | 13.16 | 02:31 | 05:54 |
| 10:60:4:0 | 4.29 | 19.03 | 14.74 | 02:33 | 05:19 |
| 10:60:4:2 | 4.14 | 19.30 | 15.16 | 02:34 | 05:37 |

IOP Conf. Series: Materials Science and Engineering **1325** (2025) 012010 doi:10.1088/1757-899X/1325/1/012010

## 3.2 Mechanical properties and aging stability

Figure 2 presents the tensile properties of NR/RR composites in terms of tensile strength, elongation at break, and 100% modulus, under varying ratios of additional chemicals. The results show that the tensile properties decreased with the addition of TMQ due to defects introduced into the RR matrix by TMQ's flake- and spherical-like particles, as indicated elsewhere [11,12]. These defects, attributed to inhomogeneity at the interface between the rubber matrix and additional solid chemicals, can propagate under applied force, as illustrated in Figures 2(A), (B), and (C). However, after aging, both tensile strength and elongation at break exhibited significant resistance to thermo-oxidative degradation under the selected aging conditions. This indicates that while TMQ reduces tensile performance and the reinforcement degree of fillers, it enhances the thermal stability of the composites. Regarding the effect of CB:MBTS ratios, increasing MBTS beyond 3 phr did not significantly improve tensile properties or thermal resistance against decomposition and degradation. This suggests that the elevated crosslink density from higher MBTS concentrations does not contribute to mechanical enhancement in the RR composites. However, formulations with HSR:CB:MBTS:TMQ ratios of 10:60:3:0-2 and 10:60:4:0-2 demonstrated the highest tensile strength and elongation at break among the composites. Conversely, the formulation with 80 phr CB exhibited the highest modulus due to the reinforcement degrees provided by the hard particulate solid filler combination.

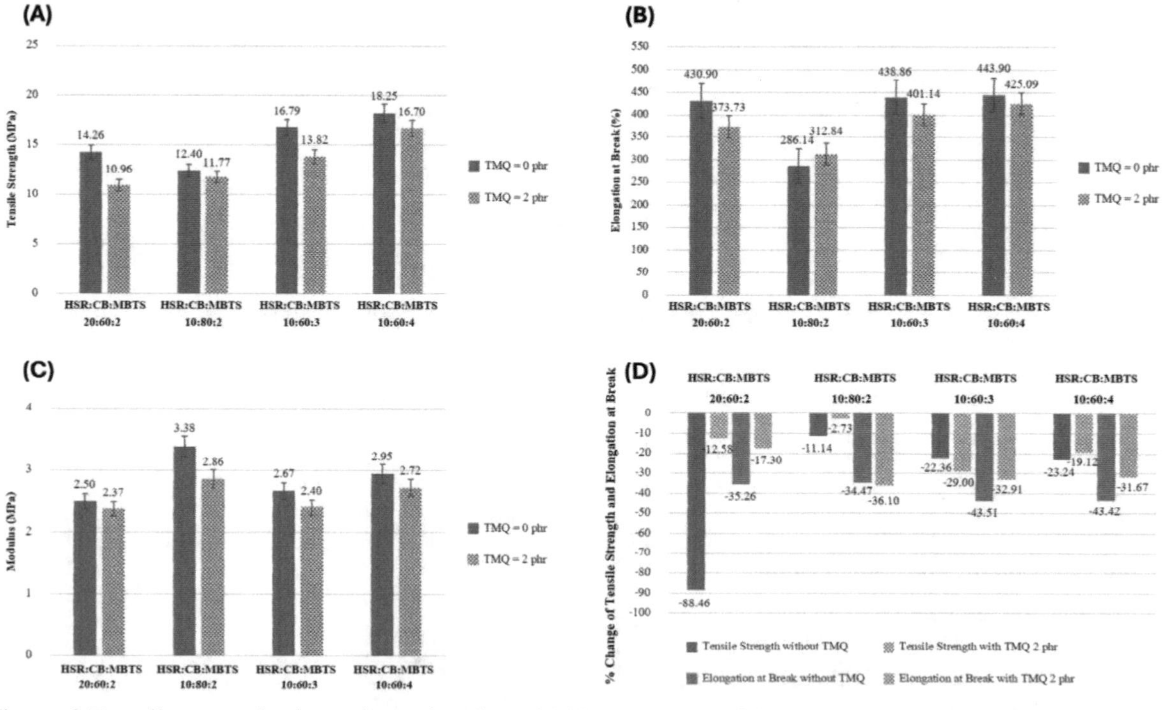


**Figure 2** Ttensile strength, elongation at break, and 100% modulus of the NR/RR composites filled with different ratios of HSR/CB/MBTS/TMQ at before and after aging.

The addition of TMQ and varying ratios of CB/MBTS not only influenced the tensile properties but also impacted tear strength and recovery properties (set properties), reflecting the material's ability to return to its original shape after tensile and compressive deformation. As shown in Table 3, tear strength exhibited no significant differences among the formulations with and without TMQ, while hardness increased, and set properties demonstrated improved recovery, achieving higher recoverability. It is seen that the HSR/CB/MBTS ratios of 10:60:4 phr indicated the highest tear strength followed by 10:60:3 phr and 10:60 and 80:2 phr. Although these results seem counterintuitive, they strongly correlate with crosslink density, which increases with higher MBTS concentrations in the composites [13]. While

IOP Conf. Series: Materials Science and Engineering **1325** (2025) 012010 doi:10.1088/1757-899X/1325/1/012010

the incorporation of TMQ introduces defects, the enhanced crosslink network among rubber molecules facilitates easier recovery of the composites to their original shape, owing to the free space within the bulk specimens. This phenomenon also explains the reduced hardness of RR composites when TMQ is added alongside high MBTS concentrations. However, the addition of TMQ with a CB:MBTS ratio of 60:3 phr enhances hardness stability during aging, as evidenced by resistance to hardness changes under conditions of 100°C for 70 h and 150°C for 24 h, as shown in Table 4.

**Table 3** Tear strenght, hardness and set properties based on tension and compression modes of the NR/RR composites filled with different ratios of HSR/CB/MBTS/TMQ at before and after aging, where the a*, b* and c* referred to no significant values from T-test.

| HSR:CB:MBTS:TMQ (phr) | Tear Strength (kN/m) | Hardness (Shore A) | Tension Set (%) | Compression Set (%) |
|---|---|---|---|---|
| 20:60:2:0 | $43.92 \pm 4.32^{a*}$ | 66 | 11 | 42.03 |
| 20:60:2:2 | $44.88 \pm 2.80^{a*}$ | 64 | 10 | 42.66 |
| 10:80:2:0 | $42.80 \pm 4.41^{a*}$ | 67 | 10 | 36.93 |
| 10:80:2:2 | $41.43 \pm 2.38^{a*}$ | 66 | 11 | 36.08 |
| 10:60:3:0 | $48.32 \pm 4.65^{b*}$ | 65 | 9 | 36.83 |
| 10:60:3:2 | $48.31 \pm 6.53^{b*}$ | 64 | 7 | 32.60 |
| 10:60:4:0 | $57.75 \pm 5.14^{c*}$ | 65 | 8 | 27.95 |
| 10:60:4:2 | $56.14 \pm 0.75^{c*}$ | 65 | 7 | 29.20 |

**Table 4** Hardness changes of the NR/RR composites filled with different ratios of HSR/CB/MBTS/TMQ after aging.

| HSR:CB:MBTS:TMQ (phr) | Hardness change (Shore A) 100°C for 70 h. | Hardness change (Shore A) 150°C for 24 h. |
|---|---|---|
| 20:60:2:0 | 5 | 6 |
| 20:60:2:2 | 4 | 6 |
| 10:80:2:0 | 4 | 8 |
| 10:80:2:2 | 4 | 7 |
| 10:60:3:0 | 4 | 3 |
| 10:60:3:2 | 3 | 2 |
| 10:60:4:0 | 5 | 3 |
| 10:60:4:2 | 5 | 2 |

### 3.3 QUV resistances

The weathering resistance test using a Fluorescent UV Weathering Tester was performed according to ASTM G154. The testing cycle consisted of UV exposure at intensities of 0.89 and 1.55 W/m² under a constant temperature of 60°C for 8 h, followed by dark conditions at a constant temperature of 50°C for 4 h, alternating for a total duration of 60 h. The relative humidity was maintained at 50 ± 5%, and tensile testing was utilized to assess the outcomes. Figure 3 illustrates the tensile properties before and after exposure to QUV conditions. Interestingly, results differed from those observed under heat aging. QUV exposure effectively increased the tensile strength and 100% modulus of the RR composites, particularly with the addition of TMQ and increased MBTS concentrations. However, there were no significant differences in elongation at break. These findings indicate that the applied UV conditions provided sufficient energy to generate radicals on the rubber chains, enhancing the chemical crosslink density of the RR composites. This effect was especially pronounced in formulations with higher MBTS concentrations, reaching an optimal level at 3 phr.

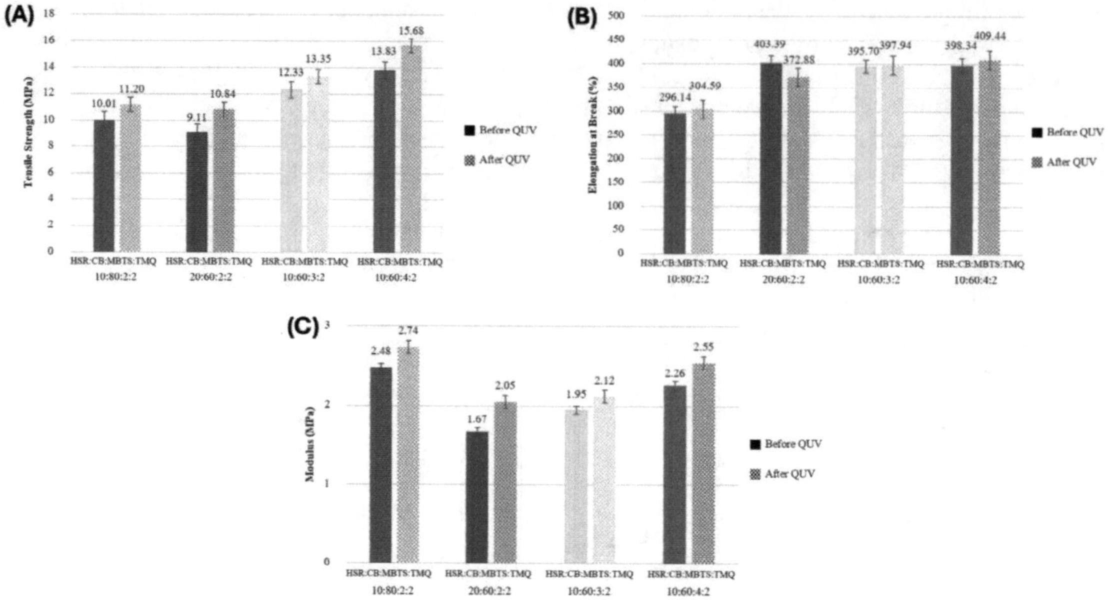


**Figure 3** Tensile strength, elongation at break, and 100% modulus of the NR/RR composites filled with different ratios of HSR/CB/MBTS/TMQ at before and after QUV applying.

## 4. Conclusions

The study demonstrates that the interplay of HSR, TMQ, CB, and MBTS concentrations significantly influences the curing behaviour, mechanical properties, and stability of RR composites. TMQ enhances thermal and UV stability while moderating tensile performance due to defect introduction. Optimal formulations were achieved with MBTS concentrations at 3 phr and HSR:CB ratios that balance reinforcement and thermal conductivity, yielding composites with superior durability and weathering resistance. These findings support the tailored design of RR composites for advanced performance in various applications, particularly in outdoor applications requiring weather resistance.

**Acknowledgement**
This research project was supported by National Research Council of Thailand (NRCT), Grant No. 175570, and also the King Mongkut's University of Technology Thonburi, Bangkok, Thailand for the facilities.

IOP Conf. Series: Materials Science and Engineering **1325** (2025) 012010 doi:10.1088/1757-899X/1325/1/012010

## References


[1] Arayapranee W and Rempel GL 2008 *J. Appl. Polym. Sci.* **109**, 1395–1402.
[2] Hamza S 1998 *Polym. Test.* **17**, 131–137.
[3] Harper M, Tarddiff J, Haakenson D, Joandrea M and Knych M 2017 SAE *Int. J. Veh. Dyn. Stab. NVH.* **1**, 179–189.
[4] Ismail H and Jessy RS 2000 *Int. J. Polym. Mater.* **46**, 205–216.
[5] Nakaramontri Y, Nakason C, Kummerlöwe C and Vennemann N 2015 *Rubber Chem. Technol.* **88**, 199–218.
[6] Nakaramontri Y, Pichaiyut S, Wisunthorn S and Nakason C 2017 *Euro. Polym. J.* **90**, 467–484.
[7] Nayak NC and Tripathy DK 2001 *J. Elas. Plas.* **33**, 179–195.
[8] Pal K, Rajasekar R, Kang DJ, Zhang ZX, Pal SK, Das CK and Kim JK 2010a *Mater. Des.* **31**, 677–686.
[9] Pal K, Rajasekar R, Kang DJ, Zhang ZX, Pal SK, Das CK and Kim JK 2010b *Mater. Des.* **31**, 1156–1164.
[10] Pal K, Pal SK, Das CK and Kim JK 2012 *Mater. Des.* **35**, 863–872.
[11] Sripornsawat B, Saiwaria S and Nakason C 2018 *J. Waste Manag.* **79**, 638–646.
[12] Nowicki J, Jaroszewska K, Nowakowska-Bogdana E, Szmatoła M and Iłowska J 2018 *Mol. Catal.* **454,** 94–103.
[13] Julbust A, Buaksuntear K, Suethao S, Kohl P, Li Y and Smitthipong W 2024 *Chin. J. Polym. Sci.* **42**, 1835–1844.

# Effect of vegetable oil on the properties of rubber compound

**S Suethao[1,2], S Phongphanphanee[1], J Wong-ekkabut[3], H Yangthong[1] and W Smitthipong[1,2,*]**

[1] Specialized center of Rubber and Polymer Materials in agriculture and industry, Department of Materials Science, Faculty of Science, Kasetsart University, Bangkok 10900, Thailand.
[2] Hub of Talents in Natural Rubber, National Research Council of Thailand (NRCT), Bangkok 10900, Thailand.
[3] Department of Physics, Faculty of Science, Kasetsart University, Bangkok 10900, Thailand.

*Corresponding author: fsciwssm@ku.ac.th

**Abstract.** Palm oil is a type of vegetable oil that can be obtained from shell and seed of palm fruits. Unpurified palm oil from shell is called crude palm oil (CPO) and unpurified palm oil from seed is called crude palm kernel oil (CPKO). Surprisingly, there is limited literature on both CPO and CPKO in rubber compounds. Therefore, this study aimed to investigate the effects of these two types of crude palm oil on the properties of rubber compounds, exploring their potential as alternative oils to replace petroleum-based oils in rubber formulations. The rubber compound formulations with both CPO and CPKO were developed and then compared their properties, including Mooney viscosity, bound rubber, curing time, and mechanical properties. It was found that rubber compound with CPO represented higher Mooney viscosity and bound rubber. Moreover, the formulation with CPO had longer scorch time and curing time because CPO exhibited higher free fatty acid (FFA) than CPKO. The FFA can neutralize base accelerator in the rubber compound. However, there were not significant differences on mechanical properties between both crude palm oil formulations. This study displays the possibility to use both CPO and CPKO in the rubber compound depending on the rubber applications, especially sidewall truck tire.

## 1. Introduction

Natural rubber (NR) is currently an important economic crop in Thailand. Then, improving the quality of tires is essential for the rubber industry. There are consideration of environmental problems and the dangers caused by synthetic chemicals, in particular tire production [1]. For example, the use of a large amount of synthetic oil with low purity is considered a carcinogen. If high purity is desired, expensive oil must be used. Therefore, reducing the amount of various chemicals in the rubber production process is a guideline that should be given importance in rubber production in the future. This includes the use of natural alternatives as vegetable oils to replace synthetic oil [2].

Ismail H *et al.* [3-4] studied palm oil fatty acid as an activator in NR/carbon black composite. It was found that palm oil fatty acid induced the increase of vulcanization time and scorch time. This also affected the dynamic properties, higher concentration of palm oil fatty acid possessed higher elastic torque value. Menon ARR *et al.* [5] studied the stability and physico-mechanical properties of NR using phosphorylated Cashew Nutshell Liquid (PCNSL) compared to aromatic oil. It was found that the use of PCNSL resulted in maximum torque compared to unoiled and aromatic oil-treated rubber and also improved tensile properties but decreased hardness due to the cluttering effect of PCNSL.

There are few literatures review on palm oil to compare between different parts of palm tree. In this study, two types of crude palm oil (CPO and CPKO) were used in rubber compound formulations. Main properties of both crude palm oils were tested, and then the rubber compound with different types of crude palm oil was characterized such as Mooney viscosity, bound rubber, curing time and mechanical properties. This work could be useful to apply for truck tire sidewall because it consumes a lot of oil for the flexibility of tire.

## 2. Experimental

Table 1 shows the rubber compound formulations used in this work. The crude palm oils were supplied by Absolute Products Co., Ltd., reclaimed rubber by Saeng Thai Rubber Co., Ltd., and other chemicals by Thai Sin Rubber Industry Co., Ltd. The order of adding chemicals in the rubber mixing process affected the quality and efficiency of the rubber compound, adding chemicals must be done while the rubber was soft. During the rubber mastication, the rubber molecules were cut off by shear force, causing the rubber to have a lower viscosity, which took 8 min using an internal mixer. Then carbon black, activators and antioxidants were added and continued mixing for 3 min. Finally, oil was also added and then mixing for 3 min. For the next step, vulcanizing agents were added in the rubber compound using 2-roll mill for 10 min.

**Table 1.** Formulations of rubber compound.

| Formulation | CPO (phr) | CPKO (phr) |
|---|---|---|
| Natural rubber (NR) | 47.50 | 47.50 |
| Butadiene rubber (BR) | 50.00 | 50.00 |
| Reclaimed rubber | 5.00 | 5.00 |
| CPO | 17.00 | - |
| CPKO | - | 17.00 |
| Carbon black (N550) | 37.50 | 37.50 |
| Stearic acid | 0.60 | 0.60 |
| Zinc oxide | 2.00 | 2.00 |
| Accelerator | 1.47 | 1.47 |
| Sulphur | 1.47 | 1.47 |
| Wax | 1.00 | 1.00 |
| Antioxidant type I | 4.00 | 4.00 |
| Antioxidant type II | 2.00 | 2.00 |

Viscosity of crude palm oils was tested at 40 °C according to ASTM D445 standard and the percentage of free fatty acids was tested according to AOCS Official method Ca 5a-40. Mooney viscometer was used to measure Mooney viscosity of the rubber compound at 100°C for heating 1 min and measuring 4 min [1]. Bound rubber was characterized by toluene soaking method [6]. Curing time of samples was performed by Moving die rheometer at 150°C for 30 min [6]. Static (tensile test) and dynamic

IOP Conf. Series: Materials Science and Engineering **1325** (2025) 012011 doi:10.1088/1757-899X/1325/1/012011

mechanical properties (dynamic mechanical analysis or DMA) were characterized following previous study [7].

## 3. Results and discussion

The viscosity and free fatty acid (FFA) of CPO and CPKO are presented in Table 2. We found that viscosity and FFA of CPO were higher than those of CPKO, because CPO comes from the whole palm fruit, but CPKO just only comes from kernel or seed of palm fruit. When both crude palm oils were added to rubber compound (Table 3), sample with CPO had higher Mooney viscosity than that with CPKO in good agreement with the viscosity of both crude palm oils. This also affected the bound rubber of rubber compounds with different types of crude palm oil, therefore, the interaction between the rubber chains and filler in CPO was superior to that of CPKO. For the curing time, CPO compound represented higher FFA than CPKO compound, then CPO compound represented longer cure time. Because the accelerator is based on amine chemical, higher FFA neutralized amine chemical leading to longer time.

**Table 2.** Viscosity and free fatty acid as palmitic acid of CPO and CPKO.

| Testing | CPO | CPKO |
|---|---|---|
| Viscosity @ 40°C (cSt) | 40.89 | 30.37 |
| Free fatty acid as palmitic acid (FFA) (%) | 5.55 | 1.49 |

**Table 3.** Green properties of rubber compounds with CPO and CPKO.

| Formulation | CPO | CPKO |
|---|---|---|
| Mooney viscosity | 15.36 | 11.03 |
| Bound rubber (%) | 46.83 | 40.04 |
| Scorch time (min) | 4.24 | 4.13 |
| Curing time (min) | 6.75 | 6.23 |

There were no significant differences in tensile at 100% and 300% for CPO and CPKO compounds (Table 4), also similar results of tensile strength and elongation at break, indicating that these compounds had similar strength (both before and after aging samples). There were also no significant differences in storage modulus (E′) and dissipation energy (tan delta) for CPO and CPKO compounds (Figure 1), these dynamic results were in good agreement with those of tensile test. This means that both types of crude palm oil in this study had no effect on the mechanical properties of rubber compound (tensile test as static mechanical properties and DMA as dynamic mechanical properties).

**Table 4.** Tensile test of rubber compounds with CPO and CPKO.

| Testing | CPO | | CPKO | |
|---|---|---|---|---|
| | - | Aging | - | Aging |
| Tensile 100% (MPa) | 1.7 | 2.2 | 1.5 | 2.3 |
| Tensile 300% (MPa) | 4.9 | 6.3 | 4.2 | 6.2 |
| Tensile strength (MPa) | 10.4 | 9.1 | 10.2 | 9.5 |
| Elongation at break (%) | 568 | 428 | 589 | 438 |

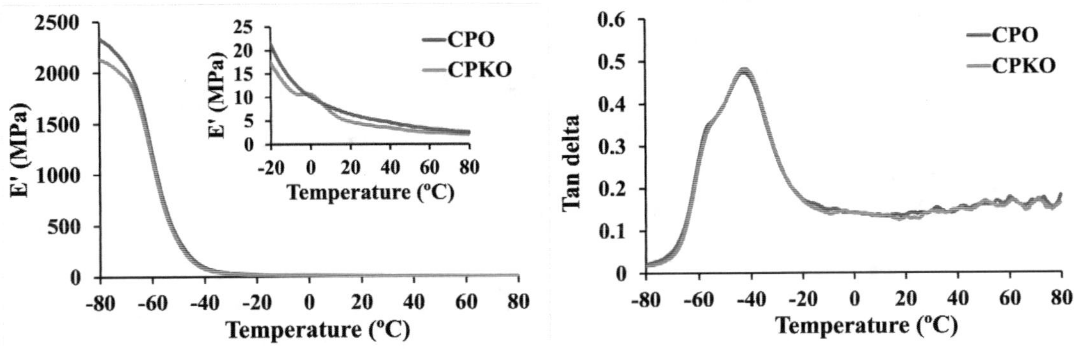


**Figure 1**. DMA of CPO and CPKO compounds: (left) storage modulus (E′) and (right) tan delta.

## 4. Conclusions

Both types of CPO and CPKO were studied in rubber compounds. CPO represented higher viscosity and FFA, that's why rubber compound with CPO had higher Mooney viscosity, bound rubber and cure time, certainly due to neutralization between FFA and amine accelerator. However, rubber compound with either CPO or CPKO had similar static (tensile test) and dynamic (DMA) mechanical properties. This research should be used for the development of eco-friendly sidewall truck tires compound with high concentration of oil.

## Acknowledgement

This research was financially supported by the National Research Council of Thailand (NRCT), Thailand. This research was also supported by the Specialized center of Rubber and Polymer Materials in agriculture and industry (RPM), Department of Materials Science, Faculty of Science, Kasetsart University, Bangkok, Thailand.

## References

[1] Promhuad K, Prasopdee T and Smitthipong W 2021 *SPE Polymers.* **2** 172.
[2] Siwarote B, Sae-Oui P, Wirasate S and Suchiva K 2017 *J. Rubber Res.* **20** 1.
[3] Ismail H, Anuar H and Tsukahara Y 1999 *Polym. Int.* **48** 607.
[4] Ismail H and Anuar H 2000 *Polym. Test.* **19** 349.
[5] Menon ARR, Pillai CKS and Nando GB 1999 *J. Appl. Polym. Sci.* **73** 813.
[6] Julbust A, Buaksuntear K, Suethao S, Kohl P, Li Y and Smitthipong W Chinese 2024 *J. Polym. Sci.* **42** 1835.
[7] Buaksuntear K, Panmanee K, Wongphul K,Lim-arun P, Jansinak S, Shah DU and Smitthipong W 2024 *Polymers.* **291** 126626.

# AUTHOR INDEX


Boochatum, P. .......... 46
Chaipakdee, N .......... 9
Chaiwichian, N .......... 5
Dechwayukul, C .......... 1
Fudulwatjananon, T. .......... 13
Hancharoen, K .......... 41
Hemvichian, K .......... 5, 9
Jaisabai, P .......... 17
Jitchati, R .......... 17
Kaewpradit, P .......... 1
Kanemaru, Kodai .......... 37
Keawwattana, W .......... 41
Ketnapha, B .......... 25
Khunkayan, Supansa .......... 32
Kittiya, W .......... 25
Klinjan, Siriwan .......... 32
Lertsarawut, P .......... 5, 9
Mingvanish, W .......... 46
Nakano, Haruyuki .......... 37
Nakaramontri, Y .......... 46
Nalakarn, P. .......... 13
Nisoh, N. .......... 13
Paeprapa, H .......... 41
Peasuwan, V .......... 41
Phongphanphanee, S .......... 54
Raoarun, M .......... 41
Rieagnam, P .......... 17
Rimdusit, R .......... 5
Rimdusit, S .......... 9
Rukkiat, J .......... 25
Saranyakan, F .......... 25
Shiroshita, Keisuke .......... 37
Smitthipong, W .......... 54
Smitthipong, Wirasak .......... 32
Suethao, S .......... 54
Suethao, Supitta .......... 32
Suwanpimolkul, Peerut .......... 32
Tiptipakorn, S .......... 5, 9
Wong-Ekkabut, J. .......... 13
Wong-Ekkabut, J .......... 54
Wongkhan, K .......... 17
Yangthong, H .......... 54
Yoshida, Norio .......... 37

**Institute of Physics**
Dirac House, Temple Back
Bristol BS1 6BE UK

ISSN: 1757-8981
ISBN 979-8-3313-1578-8

CURRAN ASSOCIATES INC.
proceedings.com

# International Conference on Mathematical Physics and Computational Simulation (CONF-MPCS 2023)

Journal of Physics: Conference Series Volume 2669

Oxford, United Kingdom
12-18 August 2023

IOP
Institute of Physics